Events and Well-being

I0890935

This book is the first to take an in-depth examination of events and well-being, adopting a much-needed critical approach to the study of events. It uses empirical case studies to help us better understand how events foster positive well-being or counter negative well-being for event organisers, participants, spectators, volunteers and even non-attending local residents.

While researchers have long understood socialisation as the major motivation to attend contemporary festivals and events, it is only just being acknowledged that well-being is also a key motivator. Those researching in the field of event studies are yet to clearly articulate "the how, why, where, and impacts of socialisation." This multidisciplinary book draws together empirical research across a range of event types and sizes, from music festivals to mega sports events, to provide a nuanced understanding of their contribution to the well-being of individuals and communities. Case studies are drawn from around the world and apply a diverse range of theoretical lenses to the conceptualisation of well-being as it applies to events and methodologies used to achieve research aims and objectives.

This significant volume will be valuable reading for students and academics in the fields of sport studies, critical event studies, queer studies, cultural studies, tourism, music, sociology and end-of-life studies.

Allan Stewart Jepson has contributed widely to event studies literature within the realm of community festivals and events. His seminal work investigated power, hegemony and the construction, representation and consumption of culture(s) at a community festival and was the first to reveal marginalised local communities trapped and underrepresented in a community cultural festival. He has three key texts in event studies, all edited with Alan Clarke, University of Pannonia, Hungary. He, along with colleagues, is currently pursuing a research agenda investigating the well-being impacts of local community festivals and events which has centred on families and Quality of Life and more recently to investigate arts participation events and collective memory creation amongst the over 70s and how these events can reduce the common psychosocial impacts associated with older age.

Trudie Walters is a freelance researcher and consultant living in Dunedin, New Zealand. Her research platform centres on events and leisure as interdisciplinary lenses through which to understand the inner workings and values of society. She is developing and managing the data strategy for Tourism New Zealand's Conference Impact programme, which seeks to create genuine, long-lasting social legacies from conferences and is a consultant with TRC Tourism. She serves as President of the Australian and New Zealand Association for Leisure Studies. In her spare time, she runs, cooks, photographs, potters about in her garden, explores new places with her husband, panders to the needs of her cat and fossicks about in second-hand shops of all kinds.

Routledge Critical Event Studies Research Series
Edited by Rebecca Finkel,
Queen Margaret University, UK
and David McGillivray,
University of the West of Scotland, UK

Gendered Violence at International Festivals
An Interdisciplinary Perspective
Edited by Louise Platt and Rebecca Finkel

Events and Well-being
Edited by Allan Stewart Jepson and Trudie Walters

For more information about this series, please visit: www.routledge.com/tourism/series/RCE

Events and Well-being

Edited by Allan Stewart Jepson
and Trudie Walters

Routledge
Taylor & Francis Group
LONDON AND NEW YORK

First published 2021
by Routledge
2 Park Square, Milton Park, Abingdon, Oxon OX14 4RN

and by Routledge
605 Third Avenue, New York, NY 10158

Routledge is an imprint of the Taylor & Francis Group, an informa business

British Library Cataloguing-in-Publication Data
A catalogue record for this book is available from the British Library

Library of Congress Cataloging-in-Publication Data
A catalog record for this book has been requested

ISBN: 978-0-367-49180-2 (hbk)
ISBN: 978-1-032-01176-9 (pbk)
ISBN: 978-1-003-04493-2 (ebk)

Typeset in Times New Roman
by Apex CoVantage, LLC

Allan:

I dedicate this book to the memory of Susan Joan Jepson (30 September 1948–25 December 2020), my mum, and my best teacher

Because of you I learnt to share

Because of you I learnt to love unconditionally

Because of you I treat everybody equally and with respect

Because of you I learnt to read and to critique what I had read, and to search for the truth

Because of you I became a good teacher

I am who I am and what I am, because of you mum

You will always live on in my heart, and I will carry you with me always

Thank you to my friend Trudie for putting up with me for a second important book and for your dedication and professionalism throughout. This is another important addition to event studies and there is nobody else I would have chosen to work with on this project.

Thank you to all of our contributing authors for coming on this journey with us and for presenting your research to the highest standard and also to those who have contributed anonymously to this book in order that your voices are heard, please trust us when we say they will be.

Trudie:

This book is dedicated to my husband Richard – thank you for your love and support over the last 27 years and for encouraging me to expand my horizons and move into a place I can flourish. May we never stop having adventures!

I would also like to thank Allan for suggesting we work together again; it has been a pleasure to help bring another high-quality body of work to light. And in the process, a valued second collaboration has well and truly cemented a valued friendship. I look forward to our next project.

Lastly, special thanks must go to the contributing authors for trusting us and for the insightful research you have done/are continuing to do in pushing the boundaries of critical event studies. It has been a joy to work with you all.

Contents

Figures

Tables

Contributors

Rachel J. Batty is Senior Lecturer within the School of Sport, Exercise and Nutrition at Massey University (Auckland, New Zealand). Her research interests include managing sport stakeholder relationships, sport organisation leveraging opportunities and strategic approaches to sport sponsorship and also skilled in qualitative research design and analysis. She has extensive experience in undergraduate/postgraduate teaching within the sport discipline, including the coordination of courses focused on sport events, sport marketing, sport development and sport business management. She is the Sport Development discipline leader for Massey's Bachelor of Sport and Exercise degree and also works externally alongside a range of sport industry providers.

Kerri Bodin is a PhD candidate in the Faculty of Health Sciences at the University of Ottawa (Ontario, Canada). Her research focuses on social and political aspects of sport events and the impacts of events on host community members. Her doctoral research investigates how residents' experiences with sport events may influence their relationship with the local government. She employs qualitative and quantitative approaches to investigate diverse perspectives within the sport event context.

Ela Brolich is a graduate in Tourism Marketing at the University of Rijeka, Faculty of Tourism and Hospitality Management (Croatia). Currently working as an administrative secretary, her current research interest focuses on event management and a better understanding of the link between tourism and quality of life of the local residents.

Alba Colombo, PhD, is Senior Lecturer and researcher at the Department of Arts and Humanities faculty of the Open University of Catalunya based in Barcelona, Spain. She has long-term experience in events research and education. Currently, she is a principal investigator of the Barcelona team at the HERA financed FESTSPACE research project. Her main research area is a sociology of culture, on a frame of critical event studies, focusing on the analysis of festivals and festivities as contemporary social and cultural expressions, during uncertain times. In this frame, her recent publications revolve around social effects and impacts of festivals and festivities.

Paul Fallon is Senior Lecturer in the Tourism, Hospitality and Events Division at the University of Central Lancashire (UCLan), where he is responsible for teaching at UG and PG levels, PhD supervision and managing its international partners. He has 20+ years' experience in teaching and managing projects at four UK HEIs and their international partners. During this period, he has also been an active researcher. More recently, he has extended his practice to qualitative approaches and is finding rich opportunities within the events sector to explore his interests in consumer behaviour and experiences.

Francesc Fusté-Forné is a professor and researcher at the University of Girona and a collaborating professor and tutor at the Open University of Catalonia. His research is focused on rural and food tourism and food media. Particularly, he has studied the food tourism phenomenon by bringing connections between authenticity, food heritages, landscapes and landscapers, regional development, rural activities and tourist experiences.

Troy D. Glover is Chair and Professor in the Department of Recreation and Leisure Studies and Director of the Healthy Communities Research Network at the University of Waterloo in Canada. His research focuses on transformative placemaking, the aspirational effort to transform public spaces into meaningful places of social connection, and social capital, the consequence of investment in and cultivation of social relationships allowing individuals access to resources that would otherwise be unavailable to them.

Sandra Goh is Senior Lecturer at Auckland University of Technology, Faculty of Culture and Society, in New Zealand. Her research interests follow her career trajectory including events and festival management and tourism, event travel careers, serious leisure, social worlds, transformational travel, critical event studies, placemaking and creative methodologies.

Alex W. Grebenar loves music and loves events. Combining the two as Lecturer in Event Management at the University of Central Lancashire and active in music events research. Alex has a passionate focus on the unique experience of the individual within the eventscape. Alex has also written in the field of dark tourism and teaches on a variety of visitor economy modules. Alex is a former ski *seasonnaire*, a sometime event planner, a long-time musician and forever lost in music.

Christian H. Hanser is a PhD student researching the intersections between Teacher Education, Outdoor Education and Community Development for Initial Teacher Education at the University of Edinburgh. Over a decade, he has built up his own vagabond educational practice: mobile shepherd's huts as pop-up public pedagogies. The tiny houses have been used across Europe in participatory tourism, the arts, social work and the health-care sector as mobile hubs for community-building. He trained in humanistic-existential counselling in Germany, motivated to facilitate informal and psychosocial rather than clinical settings for life stories.

Allan Stewart Jepson, PhD, has contributed widely to event studies literature within the realm of community festivals and events. His seminal work investigated power, hegemony and the construction, representation and consumption of culture(s) at a community festival and was the first to reveal marginalised local communities trapped and underrepresented in a community cultural festival. He has three key texts in event studies, all edited with Alan Clarke, University of Pannonia, Hungary. He, along with colleagues, is currently pursuing a research agenda investigating the well-being impacts of local community festivals and events which has centred on families and Quality of Life and more recently to investigate arts participation events and collective memory creation amongst the over 70s and how these events can reduce the common psychosocial impacts associated with older age.

Phillip Joy is Assistant Professor of Applied Human Nutrition at Mount Saint Vincent University, Halifax, Nova Scotia, Canada. He is a registered dietitian. His research interests include the nutritional health of LGBTQ2SP+ communities and often uses arts-based methodologies.

Jordi Oliva is a university researcher at the University of Leipzig and in the Open University of Catalonia, production manager in an opera company in Saint Sebastian (Spain) and promoting manager of several artists. His research fields are the sociology and psychology of music. He has analysed festivals in Spain and Germany. He has a degree in industrial engineering, a master's degree in cultural management and has been a visiting scholar at the University of Chemnitz (Germany). He has been a reviewer for various journals, including the *Journal of Policy Research in Tourism, Leisure and Events*.

Amanda Payne is a course leader in Event Management at the University of Central Lancashire (UCLan). Prior to lecturing, Amanda had an extensive 15-year career in the events industry planning, organising and delivering numerous national and international art events. Ultimately, she is committed to embedding "real world" context into her teaching and research. She is also currently undertaking a PhD, the focus of which investigates whether art and cultural tourism/events provision can aid those living in poverty.

Marko Perić is Associate Professor at the University of Rijeka, Faculty of Tourism and Hospitality Management (Croatia). His research and teaching activities have concentrated on the area of event management, strategic management and sport management issues. Main results of his work have been published in prestigious international journals such as *Journal of Sport & Tourism* amongst many others.

Maria Pont-Llagostera graduated in Tourism at the University of Girona. She has experience in tourism management in both public and private organisations. Specially, she is involved with destination and event planning and the promotion of culinary heritages. Her research interests include gastronomy and wine tourism, sustainable management and language learning.

Stephen Pratt is Professor at the School of Tourism & Hospitality Management, The University of the South Pacific. He was appointed to this position in September 2018. Previously, he was Assistant Professor at the School of Hotel & Tourism Management at The Hong Kong Polytechnic University from January 2013. He completed his PhD at the University of Nottingham, UK. He obtained a Bachelor of Economics and a Master of Economics from the University of Sydney. His research interests include sustainable tourism development (economic, socio-cultural and environmental impacts of tourism), tourism in small island states and tourist behaviour.

Nacanieli Rika is Senior Lecturer in the School of Accounting and Finance at the University of the South Pacific. He holds a PhD from the Australian National University. In 2018, he completed a fellowship at the University of New South Wales as a titular fellow of the Association of Commonwealth Universities. His publications focus on accountability, social and environmental accounting and public sector accounting, with particular emphasis on Pacific Island countries. In 2020, he was the joint recipient of the John Perrin Prize for Best Paper.

Claudio M. Rocha is Lecturer in Sport Management at the University of Stirling, UK. He previously worked as an assistant professor at the University of São Paulo (USP), Brazil, a visiting scholar at the University of Massachusetts Amherst and a Fulbright Scholar at Ohio State University, USA. He has a PhD in sport management from Ohio State University and an MPhil in Sport Studies from the University of São Paulo. His research interests are in the social impacts of sport mega-events.

Jagdeep Kaur Sabharwal is an educational professional with an expertise in Stress Management and Mindfulness training. She has worked for many years as a corporate trainer before she shifted her career and took up teaching and research at the James Cook University. Her research focuses on the social, relational and cultural influences on well-being, stress management and mindfulness, burnout, psychosocial gerontology and intergroup processes. She has a special interest in e-mental health applications to help reduce stress and increase well-being.

Ismail Shaheer is Postdoctoral Fellow in the Department of Tourism, University of Otago. His research focuses on tourist behaviour, ethics, social media and social movements. Before undertaking his doctorate at the University of Otago in New Zealand, he worked at the Maldives Tourism Promotion Board (now Maldives Marketing & Public Relations Corporation) for more than ten years marketing the Maldives as a tourist destination.

Shavneet Sharma was born in Fiji. He received the BCom degree in accounting and management and the MCom degree in 2015 and 2018, respectively, from the University of the South Pacific, Fiji, where he is currently working towards the PhD degree in marketing with the School of Management and Public Administration. He was awarded the Gold Medal and Vice-Chancellor and President's Prize for the most outstanding graduate with a Management and Public Administration Major. In 2016, he joined the University of the South Pacific as a teaching assistant. His research interests include information systems, technology adoption, digital marketing and social media marketing.

Amrit Singh is a funeral arranger for one of the largest funeral providers in the UK. She takes care of the clients from the moment they make the first contact until the client no longer requires support. Before dedicating her work to the bereaved, she secured a first-class BA (Hons) degree in Event Management at the University of Central Lancashire (UCLan) in 2019, where she was able to extensively develop an interest in the funeral industry. She was also awarded the "Inspiring Female Student award" that showcased and illuminated enterprising female students and their experiences, impactful research and best practice.

Marijke Taks, PhD, is Vice-Dean Research and Full Professor in the Faculty of Health Sciences at the University of Ottawa (Ontario, Canada). She is also adjunct professor in the Department of Kinesiology at the University of Leuven (Belgium). Her area of expertise is in socio-economic aspects of sport and leisure. Her grant-supported research focuses particularly on impacts, outcomes and leveraging of sport events and their meaning for host communities. She is a past editor of the *European Sport Management Quarterly* and an associate editor for various sport management-related journals. She is an accomplished teacher in the political economy of sport, sport management and marketing.

K. Thirumaran is Academic Head of James Cook University Singapore Business School. Thiru's research interests are in cultural, transformational and luxury tourism.

Trudie Walters is a freelance researcher and consultant living in Dunedin, New Zealand. Her academic research platform centres on events and leisure as interdisciplinary lenses through which to understand the inner workings and values of society. Her industry work includes the development of a data strategy for the Tourism New Zealand Conference Impact Aotearoa programme, which seeks to create genuine, long-lasting social legacies from conferences. She is President of the Australian and New Zealand Association for Leisure Studies.

Nicholas Wise is Assistant Professor in Tourism Planning and Development at Arizona State University, his work focuses on cultural events and event impacts, with recent work focuses on destination development and the European Capital of Culture. He has edited nine books and has published recent papers in reputable journals.

1 Conceptualising the intersections of events and well-being

Allan Stewart Jepson and Trudie Walters

Introduction

In developing this book, we ran the risk of being seen to 'jump on the well-being bandwagon', as the interest in well-being has increased across a diverse range of discipline areas in recent years such as music and the arts (see, e.g. Packer & Ballantyne, 2011), sport (Lee, Cornwell & Babiak, 2012), tourism (Smith & Diekmann, 2017), economics (McDaid & Cooper, 2014), interior architecture (Smith, Metcalfe & Lommerse, 2012) and of course the health sciences, including psychology (Jetten, Haslam & Haslam, 2012). However, the foundations of this book were laid through doctoral study (Jepson, 2009) that investigated the non-inclusiveness of local communities within festival planning processes through hegemonic relationships. More recent origins thought lie in our previous edited volume, *Marginalisation and Events* (Walters & Jepson, 2019), as many of the contributors made explicit links between events and concepts such as quality of life, identity, community cohesion, empowerment, resistance and transformation for the marginalised groups and communities whose stories were being told. We concluded that events could indeed help resist or overcome marginalisation and thus contribute to well-being. The idea for *Events and Well-being* was born.

Again, our aim for this volume was to bring together a multidisciplinary collection of work with robust primary data and analysis, which applied a critical approach to events as they relate to well-being that delved deeper and made use of a range of different theoretical and methodological approaches to provide a nuanced understanding of the intersection of events and well-being. The contributing authors have engaged with these themes enthusiastically and critically, and this book will appeal to both emerging and established scholars working in a range of discipline areas and fields of study – from sport to critical event studies, from queer studies to cultural studies and from end-of-life to music. Our edited collection also holds value for industry practitioners, with many insightful recommendations to foster well-being provided throughout the chapters.

This first chapter seeks to help the reader conceptualise the intersections of events and well-being and begins with a definition of 'well-being'. This provides the necessary context for an examination of events and the ways in which they may engage with well-being, and the chapter then draws out the contributions of subsequent chapters.

What do we understand by 'well-being'?

There are multiple ways to understand and interpret the term 'well-being' including physical, psychological and physiological (Packer & Ballantyne, 2011; Dunn, 2014; Stebbins, 2018), but the chapters in this volume mirror the existing work in critical event studies and almost unanimously utilise subjective well-being as the most appropriate measurement of the intersection of events and well-being (Packer & Ballantyne, 2011; Jepson, Stadler & Spencer, 2019). Yang (2008, p. 204) defines subjective well-being as 'a state of stable, global judgement of life quality and the degree to which people evaluate the overall quality of their lives positively', and it thus relies on how an individual perceives and evaluates their life (Diener, 1984). Perhaps unsurprisingly, satisfaction with life, overall happiness, and enjoying one's work (if employed) all contribute to a higher level of well-being, whereas stress, anxiety and depression are associated with lower levels of well-being (Steger *et al.*, 2006). Davidson and Cotter (1991) consider subjective well-being to be comprised of three components: positive affects (such as feelings of happiness, excitement, being pleased and cheerfulness), negative affects (including being free of sadness, guilt, anger and excessive worry) and perceived efficacy or being able to handle one's life.

Well-being thus encompasses physical, physiological, sociological and psychological elements. An important message of warning is emerging in the 21st century, one to which researchers in the fields of tourism, hospitality and events have largely been apathetic: that sociological and psychological well-being are inseparable and crucial to our happiness and our survival. Research has demonstrated that social relationships are as important to mortality as quitting smoking, keeping within body mass index (BMI), lowering blood pressure and keeping physically active (Holt-Lunstad, Smith & Layton, 2010). Further empirical studies suggest that one increases their life expectancy by up to 50% if adequate social relationships are in place (House, Umberson & Landis, 1988). Researchers within contemporary leisure industries such as tourism, hospitality and events are well placed to conduct research into understanding the intricacies and importance of the experiences that these industries create and the impact upon human sociological, physiological and psychological health and well-being.

Within events studies, there is a growing base of literature seeking to better understand how events can enhance well-being (Clift & Hancox, 2010; Bosnjak *et al.*, 2016; Armbrecht & Andersson, 2019), through solicited emotions (Wood & Moss, 2015), or emotional togetherness (Wood & Dashper, 2020). Research in well-being has focused upon subjective inquiry such as human needs, human growth and flourishing (Seligman, 2011) and hedonic and eudemonic experiences (Voigt, Howat & Brown, 2010). It has also been explored in more specific contexts such as well-being in the over 70s (Stadler, Jepson & Wood, 2020; Jepson, 2019; Wood, Jepson & Stadler, 2018) within families (Jepson, Stadler & Spencer, 2019; Stadler & Jepson, 2018) or to understand how positive power relationships within communities can leverage well-being by creating more inclusive events and thus de-marginalising communities (Walters, Stadler & Jepson, forthcoming).

While researchers have long understood socialisation as the major motivation to attend contemporary festivals and events, it is only just being acknowledged that well-being is also a key motivator (Weinberg & Joseph, 2017). Researchers in the field of event studies are yet to clearly articulate 'the how, why, where, and impacts of socialisation'. If these aspects are understood, they are only from a neurotypical audience, so there are clear opportunities for researchers to understand how important events might be for neurodiverse audiences, or those with complex health conditions.

Events and well-being

In recent years, several event studies scholars have investigated the intersection of subjective well-being and event attendance and participation, whether explicitly or more obliquely using terms such as 'quality of life' or 'life satisfaction'. This research includes work on aspects of well-being for refugee and ethnic minority migrants (Hassanli, Walters & Friedmann, 2019; Hassanli, Walters & Williamson, 2020; Walters & Venkatachalam, forthcoming), families (Jepson & Stadler, 2017), residents and communities (see, e.g. Gibson & Connell, 2015; Yolal *et al*., 2016; Walters, 2019), charity sporting event participants (Filo & Coghlan, 2016) and young people (Packer & Ballantyne, 2011).

A common thread throughout these studies is that events contribute to attendees' quality of life and, therefore, facilitate a sense of well-being through their very nature as fora for social connection that draw like-minded people together (Derrett, 2003). This may be due to the resultant affirmation of community identity and values, the fostering of social cohesion or the boosting of resilience in a difficult time (Derrett, 2009; Jepson & Clarke, 2013; Gibson & Connell, 2015; de Brito & Richards, 2017; Jepson & Stadler, 2017; Duffy & Mair, 2018; Walters, 2019). In a music festival context, in addition to actively celebrating music with other event attendees, the shared experience of interacting and engaging with the performers is an important well-being contributor (Packer & Ballantyne, 2011). Cultural events allow participants a freedom to be themselves and to share their culture with a wider audience in a safe environment, creating a sense of community that does not exist to the same extent in their everyday lives (Hassanli, Walters & Friedmann, 2019; Hassanli, Walters & Williamson, 2020; Walters & Venkatachalam, forthcoming).

Therefore, we argue here that multidisciplinary and/or interdisciplinary research design should be embedded into all future events studies where understanding event experience and well-being is a key research outcome alongside being part of the critical event studies movement. Psychology, sociology and marketing domains remain the key to unlocking a better understanding of festival and event experiences, memories and with it a clearer understanding of how best to leverage impacts, which ensure we continue to maintain our positive sociological, physiological and psychological well-being.

The chapters in this volume have an explicit focus on the intersection of events and well-being, and as such, they provide a nuanced understanding of how and when events may contribute to well-being for attendees. Furthermore, they cover a diverse range of event types and sizes, and stakeholder groups including event organisers, event participants, volunteers, event spectators and even local residents who did not attend the event(s). Some chapters have adopted what we may consider a 'deficit approach' whereby events are found to be a means of addressing a well-being gap or lack. Others have taken an 'enhancement approach' where the events contribute to improving well-being in some way. As such, we have divided the book into two parts based on these broad themes: (1) events and fostering positive well-being and (2) events and countering negative well-being.

Part I: events and fostering positive well-being

The first four chapters in this theme seek to investigate the value of sports events in fostering positive well-being. Certainly, the physical well-being benefits of sports participation are self-evident: reduced risk of heart disease, diabetes and other chronic conditions, improved muscle and bone health and weight control, amongst others. These benefits have also been extrapolated to sport event participation and attendance, which are said to lead to increased physical activity (through events with sport stars functioning as inspiration). Subjective well-being is also held to result from sport event participation through instilling, for example, feelings of national pride. However, findings from existing research have not been conclusive, and it is therefore unclear whether sport events improve subjective well-being, how long such outcomes may last, or through which mechanisms these outcomes are possible. The four chapters leading Part I all attempt to address these issues with a focus on community.

In Chapter 2, Bodin, Rocha and Taks use the 2016 Rio Olympic Games as a case study to interrogate the subjective well-being and social experiences of non-attendee host country residents, as justifications for hosting sport mega-events are often based on their supposed ability to generate positive social outcomes for the residents of the host country. Their research uses a multi-item measure to investigate social impact experiences as potential mechanisms for subjective well-being for residents of Brazil during and six months after the Games were held. They find that evaluations of subjective well-being do not change between the two points in time, and thus they conclude that there is little support for the assertion that hosting the Olympic Games can change subjective well-being for residents who do not attend the event.

Batty (Chapter 3) likewise begins by discussing the more obvious physical-health outcomes of participation in sport events noted earlier, as well as the more intangible psychological well-being outcomes (such as collective identity, sense of belonging) – but notes that both are context-specific and depend on the type of engagement. Her chapter presents an exploration of the role of sport events (specifically a professional baseball tournament) in creating well-being outcomes for the community in which the home team is based, and how such events might be leveraged to facilitate maximum benefit.

The 2019 Pacific Games held in Samoa is the focus of Pratt, Rika and Sharma's contribution (Chapter 4). They start by providing an extensive list of economic and socio-cultural benefits that sport events can bring to a destination: some (such as increased sense of community pride and social cohesion) are associated with improved subjective well-being. They survey Samoan residents who attend the Games to assess whether attendance motivates them to live a healthier and more environmentally sustainable lifestyle – and find that, in general, it does, although further work is needed in the form of a follow-up survey to determine how long-lasting these well-being benefits are.

In Chapter 5, Goh, Sabharwal and Thirumaran do find significant flow-on well-being effects of a sports event – even for residents of a non-host city. Their fascinating research on the impacts of a Singaporean swimmer winning the country's first-ever gold medal, in the 2016 Rio Olympic Games, stands in sharp contrast to the lack of subjective well-being benefits for non-attendee residents during and after the same event presented by Bodin, Rocha and Taks earlier. In so doing, it supports Batty's contention that well-being outcomes of sport events are context-specific. Using netnography, themes are extracted using the PERMA framework, and quotes are used to clearly illustrate each point. They identify long-lasting meaning, accomplishment and identity as a result of the Olympic win, even four years after the event, and thus conclude that these well-being outcomes have been sustained.

We then move on to an exploration of a religious event (Advent) and its ability to foster positive well-being in Brolich, Perić and Wise's chapter, where again the attention is turned to the experiences of residents rather than visitors. Advent is Zagreb's most important event and draws thousands of tourists to the town each year, with high demand meaning the four-week event has spread from the town centre to other parts of the city to try and reduce crowding. Interestingly, their study reveals that residents feel Advent in Zagreb has a more favourable effect on the quality of life of others than on their own personal quality of life. The authors conclude by providing valuable suggestions for event managers to consider to improve the Advent offering to mitigate some of the factors that detract from residents' quality of life – and these recommendations will be of use to event organisers in other event contexts also.

In Chapter 7, Fusté-Forné and Pont-Llagostera take us on a gastronomic journey to show how a food event that celebrates a heritage bean can promote well-being. Naturally, food makes a significant contribution to our physical well-being as it nourishes and sustains us, but it also has intellectual, emotional, social, identity and symbolic aspects that may contribute to a sense of subjective well-being. Through interviews with local restaurateurs who participate in an annual Catalonian Bean Fair, this chapter finds that the event not only nurtures a sense of cultural identity and pride, but it also helps develop and maintain social networks centred around food. These factors, along with a sense of place and a focus on local lifestyles and traditions at the event, help foster life satisfaction amongst the research participants.

End-of-life events are the focus of Singh, Payne and Fallon's chapter, where the well-being of the funeral arranger is often overlooked as that of the bereaved (and, where possible, the deceased) is top of mind. Funeral arrangers need to have a set of soft skills that they can use to deal with the stress and grief of the bereaved to ensure that their well-being needs are met. The chapter culminates in the presentation of a model, which provides a valuable breakdown of the skillset, which can be developed by event managers considering the opportunities that end-of-life events offer. The model will also be a useful framework for a personal risk assessment to be created with the aim of improving positive well-being when working in end-of-life settings.

Grebenar (Chapter 9) investigates the comeback of vinyl and the surprising well-being outcomes that open decks events facilitate, albeit as by-products, amongst devotees. Such events result in a sense of community that is valued by participants and contributes to their well-being. However, personal well-being is also enhanced through the creation of a safe space in which to perform and be judged while at the same time expressing oneself through one's music choices and style. Interestingly, the vinyl records themselves also provoke well-being not only through their tangible embodiment of memory but also through the emotional response.

Music is also the subject of the final chapter in this theme. Oliva and Colombo utilise a combination of the PERMA framework, GEMS scale and evaluation of intense musical experiences (IME) to gather and analyse data related to the emotional outcomes of experiences at music festivals and relate these to the positive enhancement of well-being. They find that it is attendees' intense positive emotions and equally intense engagement with the music, along with a sense of achievement through being able to develop their cultural identity, that lead to improved subjective well-being.

Part II: events and countering negative well-being

This theme, while having just four chapters, is no less important. Indeed, some of the research in this section is highly charged with emotion as it tells stories of events raising people *out* of situations where they are experiencing negative well-being – sometimes for the first time in their life. The chapters in this section, therefore, demonstrate the potential for events to be deliberately conceived by event organisers as tools by which to counter negative well-being not just to foster positive well-being as has been evidenced in the first section.

In Chapter 11, Joy presents his research into the power of a body image art event for gay men. He argues that social and cultural norms create unique body image issues for gay men and that these issues have negative consequences for their mental and emotional well-being. As part of a wider project, he organised a community art show with other gay men (an exhibition of personal photographs of their own bodies, with accompanying stories). Participants and attendees had the opportunity to think differently about bodies and how social expectations placed on bodies can impact the health and well-being of men. The creative process leading up to and including the event, while therapeutic and emotionally healing for many participants, also provided a strong sense of belonging and connection with others that contributed positively to their well-being.

Walters and Shaheer (Chapter 12) take an entirely different approach to the intersection of events and well-being. Media constructions play a pivotal role in informing the public about things that they do not have personal knowledge or experience of – and therefore help to create the social norms referred to by Joy earlier, which may be positive or negative. This is especially pertinent for marginalised groups who are often silenced, ignored or represented as the 'other' and in such a way as to perpetuate negative stereotypes. This chapter focuses on media narratives around events for marginalised groups and finds that, for the most part, they do promote the 'normalisation' of diversity. However, except for cultural events, the media largely ignores events for the marginalised (particularly those for senior women, rural businesswomen and disabled people). Walters and Shaheer see this as an example of the media effectively silencing their voices and not bringing their positive celebrations to the attention of the public.

In an example of highly innovative event practice, Hanser interrogates the discomfort and alienation experienced by academic conference delegates, which results in a sense of negative well-being and proposes an experimental methodology to respond to this. He deploys a mobile tiny house (the 'Welcome Hut') at various conferences to provide a safe space where delegates are able to have time out, connect or disconnect and in so doing creates a form of professional self-care and coping strategy. This enables delegates to overcome the factors that prevent them from engaging meaningfully in conference activities even though they may be physically present.

In Chapter 14, the final chapter of the book, Glover presents a thought-provoking theoretical piece in which he calls for event managers to recognise the potential for events as solutions to the serious complex social problems of social isolation and loneliness. This timely chapter navigates the events space through COVID-19, arguing that events have the ability to counter negative well-being through facilitating social connection – even virtually, as we have seen in 2020 during the pandemic lockdowns. He ends by providing practical suggestions for how to create events that function as interventions for those experiencing social isolation and loneliness.

References

Armbrecht, J., & Andersson, T.D. (2019). The event experience, hedonic and eudaimonic satisfaction and subjective well-being among sport event participants. *Journal of Policy Research in Tourism, Leisure and Events*. Advance online publication. https://doi.org/10.1080/19407963.2019.1695346.

Bosnjak, M., Brown, C.A., Lee, D.J., Yu, G.B., & Sirgy, M.J. (2016). Self-expressiveness in sport tourism: Determinants and consequences. *Journal of Travel Research*. 55 (1), 125–134.

Clift, H., & Hancox, G. (2010). The significance of choral singing for sustaining psychological well-being: Findings from a survey of choristers in England, Australia and Germany, *Music and Health*. 3 (1), 79–96.

Davidson, W.B., & Cotter, P.R. (1991). The relationship between sense of community and subjective well-being: A first look. *Journal of Community Psychology*. 19, 246–253.

de Brito, M., & Richards, G. (2017). Events and placemaking. *International Journal of Event and Festival Management*. 8 (1), 2–7.

Derrett, R. (2003). Making sense of how festivals demonstrate a community's sense of place. *Event Management*. 8, 49–58.

Derrett, R. (2009). How festivals nurture resilience in regional communities. In: Ali-Knight, J., Robertson, M., Fyall, A., & Ladkin, A. (eds.) *International perspectives of festivals and events*. Oxford: Elsevier, pp. 107–124.

Diener, E. (1984). Subjective well-being. *Psychological Bulletin*. 95, 542–575.

Duffy, M., & Mair, J. (2018). Engaging the senses to explore community events. *Event Management*. 22, 49–63.

Dunn, M. (2014). Understanding athlete well-being. *Sport Health*. 32 (2), 46–47.

Filo, K., & Coghlan, A. (2016). Exploring the positive psychology domains of well-being activated through charity sport event experiences. *Event Management*. 20 (2), 181–199.

Gibson, C., & Connell, J. (2015). The role of festivals in drought-affected Australian communities. *Event Management*. 19 (4), 445–459.

Hassanli, N., Walters, T., & Friedmann, R. (2019). Can cultural festivals function as counterspaces for migrants and refugees? The case of the New Beginnings Festival in Sydney. *Leisure Studies*. Advance online publication. https://doi.org/10.1080/02614367.2019.1666296.

Hassanli, N., Walters, T., & Williamson, J. (2020). 'You feel you're not alone': How multicultural festivals foster social sustainability through multiple psychological sense of community. *Journal of Sustainable Tourism*. Advance online publication. https://doi.org/10.1080/09669582.2020.1797756.

Holt-Lunstad, J., Smith, T.B., & Layton, J.B. (2010). Social relationships and mortality risk: A meta-analytic review. *PLoS Medicine*. 7 (7), e1000316.

House, J.S., Umberson, D., & Landis, K.R. (1988). Structures and processes of social support. *Annual Review of Sociology*. 14, 293–318.

Jepson, A. (2009). *Investigating cultural relationships within the festival planning and construction process in a local community festival context*. PhD Thesis. University of Derby. https://ethos.bl.uk/OrderDetails.do?uin=uk.bl.ethos.506689.

Jepson, A. (2019). 'De-marginalising' marginalised communities: The case of participatory arts events and the over 70s in rural Hertfordshire, United Kingdom. In: Walters, T., & Jepson, A. (eds.) *Marginalisation and events*. Abingdon: Routledge, pp. 76–92.

Jepson, A., & Clarke, A. (2013). Events and community development. In: Finkel, R., McGillivray, D., McPherson, G., & Robinson, P. (eds.) *Research themes for events*. Wallingford: CAB International, pp. 6–17.

Jepson, A., & Stadler, R. (2017). Conceptualising the impact of festival and event attendance upon family quality of life (QOL). *Event Management*. 21, 47–60.

Jepson, A., Stadler, R., & Spencer, N. (2019). Making positive family memories together and improving quality-of-life through thick sociality and bonding at local community festivals and events. *Tourism Management*. 75, 34–50.

Jetten, J., Haslam, C., & Haslam, S.A. (eds.) (2012). *The social cure: Identity, health and well-being*. Hove and New York: Psychology Press.

Lee, S., Cornwell, T.B., & Babiak, K. (2012). Developing an instrument to measure the social impact of sport: Social capital, collective identities, health literacy, well-being and human capital. *Journal of Sport Management*. 27, 24–42.

McDaid, D., & Cooper, G. (eds.) (2014). *The economics of well-Being. Well-being: A complete reference guide, Volume V*. Chichester: Wiley Blackwell.

Packer, J., & Ballantyne, J. (2011). The impact of music festival attendance on young people's psychological and social well-being. *Psychology of Music*. 39 (2), 164–181.

Seligman, M.E. (2011). *Flourish*. Sydney: Random House.

Smith, D., Metcalfe, P., & Lommerse, M. (2012). Interior architecture as an agent for well-being. *Journal of the Home Economics Institute of Australia*. 19 (3), 2–9.

Smith, M.K., & Diekmann, A. (2017). Tourism and well-being. *Annals of Tourism Research*. 66, 1–13.

Stadler, R., & Jepson, A. (2018). Understanding feelings, barriers, and conflicts in festivals and events: The impact upon family QOL. In: Mair, J. (ed.) *Routledge handbook of festivals*. Abingdon: Routledge, pp. 243–253.

Stadler, R., Jepson, A., & Wood, E. (2020). The benefits of events in older life. In: Page, S., & Connell, J. (eds.) *The Routledge handbook of events*. Abingdon: Routledge, pp. 525–539.

Stebbins, R.A. (2018). Leisure and the positive psychological states. *Journal of Positive Psychology*. 13 (1), 8–17.

Steger, M.F., Frazier, P., Oishi, S., & Kaler, M. (2006). The meaning in life questionnaire: Assessing the presence of and search for meaning in life. *Journal of Counselling Psychology*. 53, 80–93.

Voigt, C., Howat, G., & Brown, G. (2010). Hedonic and eudaimonic experiences among wellness tourists: An exploratory enquiry. *Annals of Leisure Research*. 13 (3), 541–562.

Walters, T. (2019). Proud to be South D: Perceptions of a street festival in a marginalised community in New Zealand. In: Walters, T., & Jepson, A. (eds.) *Marginalisation and events*. London and New York: Routledge, pp. 36–54.

Walters, T., & Jepson, A. (eds.) (2019). *Marginalisation and events*. London and New York: Routledge.

Walters, T., Stadler, R., & Jepson, A. (forthcoming). Positive power: Events as temporary sites of power which "empower" marginalised groups. *International Journal of Contemporary Hospitality Management*.

Walters, T., & Venkatachalam, T.S. (forthcoming). The difference Diwali makes: Understanding the contribution of a cultural event to subjective well-being for ethnic minority communities. *Event Management*.

Weinberg, M.K., & Joseph, D. (2017). If you're happy and you know it: Music engagement and subjective well-being. *Psychology of Music*. 45 (2), 257–267.

Wood, E.H., & Dashper, K. (2020). "Purposeful togetherness": Theorising gender and ageing through creative events. *Journal of Sustainable Tourism*. Advance online publication. https://doi.org/10.1080/09669582.2020.1803890.

Wood, E.H., Jepson, A., & Stadler, S. (2018). Making new memories; The Potential benefits of creative arts events for the Over 70s, and the impact upon their Psychosocial well-being. *Event Management*. 22 (6).

Wood, E.H., & Moss, J. (2015). Capturing emotions: Experience sampling at live music events. *Arts and the Market*. 5 (1), 45–72.

Yang, Y. (2008). Social inequalities in happiness in the United States, 1972 to 2004: An age-period-cohort analysis. *American Sociological Review*. 73 (2), 204–226.

Yolal, M., Gursoy, D., Uysal, M., Kim, H., & Karacaoğlu, S. (2016). Impacts of festivals and events on residents' well-being. *Annals of Tourism Research*. 61, 1–18.

Part I

Events and fostering positive well-being

2 Subjective well-being and social experiences with sport mega-events

The case of Rio 2016

Kerri Bodin, Claudio M. Rocha and Marijke Taks

Introduction

Sport mega-events (i.e., the Olympic and Paralympic Games [OPG], FIFA World Cup) have a global reach and scope and often serve political ambitions (Getz, 2012). Such events are sought after and bid for under the auspices that they create positive economic, social, and political outcomes for the host nation and city (McCloy, 2009; McCloy & Thibault, 2013). However, such outcomes are often not realised, and the costs of hosting (Preuss, Andreff & Weitzmann, 2019) have been found to outweigh benefits (Késenne, 2005, 2012; Müller, 2015).

Sport mega-event supporters, in particular, have been critiqued for sport mega-events' inability to meet economic goals while the elite and high-performance sport systems benefit from such events, and the host residents are left to contribute tax money to the venture. Given the lack of economic viability of sport mega-events, there has been a shift away from focusing on economic impact towards social impact when considering the benefits such events may have for a community. This is particularly important due to host community tax contributions to publicly funded events, as these individuals deserve to receive a positive return on investment (Késenne, 2005, 2012). As such, scholars have begun investigating the notion of social impact of events more closely and have turned to the concept of a social return on investment (SROI; Arvidson *et al.*, 2013) from sport events for host residents (Taks & Rocha, 2017). Social return on investment focuses on a much broader notion of value, over and above simply economic terms (Nicholls *et al.*, 2012; Davies *et al.*, 2019). Because of their financial contribution to sport events as taxpayers, residents deserve a positive SROI from sport events.

While scholars have investigated a wide variety of social impacts in the context of sport events, subjective well-being has been investigated from the perspective of sport economists as a useful measure to determine SROI of sport mega-events (e.g., Kavetsos & Szymanski, 2010). Subjective well-being is an individual's own evaluation of their quality of life (Diener, 2000). According to Diener (2000), subjective well-being involves four components: life satisfaction, satisfaction with important domains, positive affect, and low negative affect. Scholars have investigated each of these, as well as other measures in relation to subjective well-being across contexts (Dolan, Peasgood & White, 2008).

Meanwhile, public officials have become increasingly interested in how to improve the general public's well-being, and scholars have also called for the use of indicators and measures of well-being as opposed to purely economic measures for decision-making (Diener & Seligman, 2004). This broader call aligns well with sport event research, where scholars have been increasingly interested in the notion of an SROI. Festivals and events have been seen as ways to contribute positively to a community's subjective well-being (Kavetsos & Szymanski, 2010; Yolal *et al.*, 2016); however, there remains ambiguous empirical evidence to support this claim, particularly in the context of sport events (Taks *et al.*, 2016; Taks & Rocha, 2017). While there is some evidence to suggest that hosting a sport event does affect happiness and subjective well-being (e.g., Maennig & Porsche, 2008; Kavetsos & Szymanski, 2010), there remain questions regarding the effectiveness of sport mega-events in producing positive subjective well-being outcomes, to what extent, and how these outcomes may occur.

Therefore, the purpose of this chapter is to outline an example of how subjective well-being and social experiences have been addressed in relation to a sport mega-event and to offer considerations for future research in this area. The chapter will begin with an overview of existing well-being literature in the sport (event) context, followed by a detailed case where subjective well-being and social experiences were studied in the context of the Rio 2016 Olympic Games[1]. We will then discuss considerations and future directions for this burgeoning line of inquiry.

Literature review

Theoretical background: conceptualising and measuring well-being

Well-being is a well-researched concept in both psychology and economic domains (Dolan, Peasgood & White, 2008). Subjective well-being, psychological well-being, and social well-being are distinct concepts addressing various aspects of human experiences (Keyes, 1998; Chen *et al.*, 2013). Psychological, or eudemonic, well-being refers to the fulfilment of human potential and meaning (Chen *et al.*, 2013), whereas subjective, or hedonic, well-being refers to an individual's evaluation of their own quality of life (Diener, 2000). Further, social well-being refers to an individual's perception of their contribution to, and participation in, society (Keyes, 1998). For the purposes of this chapter, we will focus on subjective well-being.

Despite a wide array of literature in the area of well-being, there is little consensus regarding components of subjective well-being and scales with which to measure each of them (Dolan, Peasgood & White, 2008). For Diener (2000), subjective well-being involves four components: life satisfaction, satisfaction with important life domains, positive affect, and low negative affect. In sport research, subjective well-being and happiness are at times used synonymously, or happiness is used as a proxy for life satisfaction or subjective well-being (e.g., Hallman, Breuer & Kühnreich, 2013), while other scholars have considered happiness as just one aspect of subjective well-being (Seligman, 2010).

Given the various ways subjective well-being has been conceptualised, there are multiple approaches for measuring subjective well-being. One approach involves a single-item measure of happiness, whereby the participant is asked to rate their general feeling of happiness on a scale. The other approach involves a holistic approach to measuring subjective well-being, thereby reflecting the notion that subjective well-being comprises various components (Diener, 2000; Diener & Seligman, 2004). While there is no one preferred way to do so or one accepted scale (see, e.g., Positive and Negative Affect Scale [PANAS], Satisfaction With Life Scale [SWLS]), many scholars suggest that a holistic approach to measuring well-being is more suitable for understanding the concept than a single-item measure (e.g., Diener & Seligman, 2004; Seligman, 2010).

In the context of sport studies, sport economists often rely on a single-item measure to investigate happiness, life satisfaction, or subjective well-being through large-scale surveys (e.g., Kavetsos & Szymanski, 2010; Dolan *et al.*, 2019). While this approach has offered significant insight into the relationship between sport and well-being, a holistic approach to measuring the concept should offer a better understanding of the complexities of subjective well-being (Littlejohn *et al.*, 2016). Recently, sport event scholars have adopted a holistic approach to studying the concept by drawing upon Diener's (2000) conceptualisation (e.g., the Sport Event Residents' Happiness Scale; Littlejohn *et al.*, 2016). This holistic approach of measuring subjective well-being assumes that sport event experience might improve the subjective well-being by different means. Event experience is complex and may take on various meanings in one's life (Getz, 2012; Getz & Page, 2016). For example, attending elite sport competitions can help people to increase positive (e.g., enjoyment) and decrease negative (e.g., frustration) life effects. Volunteering at events can make them happy, while simply talking with friends about sport events can improve one's life satisfaction. Therefore, if we want to further advance the study of subjective well-being in association with sport event experiences, a holistic approach of the concept is necessary.

Subjective well-being, sport participation, and elite sport success

The link between sport/physical activity participation and subjective well-being is well documented (e.g., Downward & Rasciute, 2011; Sato, Jordan & Funk, 2015; Downward, Hallman & Rasciute, 2018; Kim & James, 2019). Downward and Rasciute (2011) used cross-sectional survey data collected from a national survey in the United Kingdom to investigate the effect of sport participation on subjective well-being. The survey measured subjective well-being based on a single-item happiness question, where individuals were asked to rate their happiness on a scale from 1 to 10. The authors found that overall, participating in sport and physical activity had a positive effect on subjective well-being as measured this way and that this effect was stronger amongst individuals who participated in sports with a social interaction component (Downward & Rasciute, 2011). Similarly, a study completed by Downward, Hallmann and Rasciute (2018) investigating sport, health, social capital, and subjective well-being found that sport led to subjective well-being but was mediated through health. Further, the authors found that there was no relationship between social capital and sport (Downward, Hallmann & Rasciute, 2018). Some work has found that depending on the type and frequency of sport participation, individuals may experience diminishing returns on subjective well-being (Rasciute & Downward, 2010). However, Panza *et al.* (2017) found no evidence that suggests subjective well-being may decrease with increased physical activity intensity. Their results supported other findings that indicate a positive relationship between physical activity and subjective well-being (Panza *et al.*, 2017).

Governments and policymakers have reasons to promote sport and physical activity as mechanisms to increase the health and well-being of their citizens and often justify their financial contribution to sport as such (Grix & Carmichael, 2012). Similarly, governments and policymakers have promoted the hosting of sport events as a way to increase sport participation. Based on existing research outlined earlier, this could theoretically contribute to the well-being of the population. However, the notion that sport participation increases simply due to hosting a sport event has been well-researched and no effects have been found (Weed *et al.*, 2015; Mackintosh, Darko & May-Wilkins, 2016). Therefore, the link between sport events, sport participation, and well-being is not clear.

Scholars have investigated the relationship between elite sporting success and subjective well-being, based on the assumption of policymakers that elite sport success increases national pride, happiness, and well-being (Grix & Carmichael, 2012; Hallmann, Breuer & Kühnreich, 2013). Pawlowski, Downward and Rasciute (2014) studied the link between national pride following elite sporting success and well-being. In their study, Pawlowski and colleagues (2014) found no support for the notion that national pride following elite success contributes to well-being, confirming the findings of other studies that did not find a link between elite success and subjective well-being (e.g., Kavetsos & Szymanski, 2010; Dolan *et al.*, 2019). Hallman, Breuer and Kühnreich (2013) also investigated subjective well-being and national athletic success in Germany. These authors found that interest in sports and sport participation predicted happiness when German athletes succeeded. However, it is not possible to determine causality because it is known that sport participation and physical activity influence subjective well-being positively. Therefore, the effect may not be due to athletic success, it may rather be simply that those who are more physically active and involved in sport are happier in general (Hallmann, Breuer & Kühnreich, 2013).

Subjective well-being, social experiences, and sport events

Governments and policymakers claim that sport events can be an avenue to increase sport participation and national pride, both of which governments have assumed to improve well-being (Grix & Carmichael, 2012). However, as discussed earlier, sport events have neither been found to increase sport participation (Weed *et al.*, 2015), nor has national pride from elite sport success been found to significantly affect subjective well-being (Pawlowski, Downward & Rasciute,

2014). Social impact of sport events remains a salient area of inquiry due to the lack of positive economic impact observed from such events (Késenne, 2005, 2012). Clearly, positive social experiences from events may be considered as a positive SROI (Davies *et al.*, 2019).

Research regarding social impacts of sport events has included research on social cohesion, social capital, national pride, and sport participation, amongst other topics (e.g., Karadakis & Kaplanidou, 2012; Heere *et al.*, 2013; Gibson *et al.*, 2014). In the context of sport mega-events, scholars have found some evidence for positive social impacts in the short term, with limited long-lasting effects (Gibson *et al.*, 2014; Schlegel, Pfitzner & Koenigstorfer, 2017); however, results have been ambiguous at best. Research has also investigated, albeit not as often, the potential for sport events to induce negative social impacts such as disturbance of daily life (e.g., traffic, noise, construction sites) and safety concerns (Deery & Jago, 2010; Balduck, Maes & Buelens, 2011). Further, most of the research regarding the social impacts of sport events has focused on resident perceptions of social impacts and individuals' attitudes towards event impacts. That is, researchers have asked individuals how they feel about certain impacts rather than asking about their individual lived experiences. Empirical evidence based on actual lived social experiences has seldom been completed and remains an area for further investigation. Similar to the social impacts of sport events, a few studies have investigated subjective well-being as a possible lasting social effect following the event (Taks *et al.*, 2016; Taks & Rocha, 2017).

There is some research to suggest that hosting sport events increases subjective well-being in the host city and country at least during the event (Maennig & Porsche, 2008; Kavetsos & Szymanski, 2010). Maennig and Porsche (2008) asserted that the feel-good effect was the most detectable outcome of the 2006 FIFA World Cup. Kavetsos and Szymanski (2010) investigated the effect of hosting sport events and national team success on overall life satisfaction. The authors conceptualised life satisfaction as synonymous with well-being and the "feel-good factor". They used data from 12 different countries gathered via the Eurobarometer, which uses a single measure of life satisfaction stating, "On the whole are you very satisfied, fairly satisfied, not very satisfied, or not at all satisfied with the life you lead?" (Kavetsos & Szymanski, 2010, p. 161). The authors found that the feel-good factor related to hosting football events was significant, however, the effect of national team success on feel-good factor was not significant. Further, the authors observed no long-term (i.e., legacy) effects of the feel-good factor. Similarly, Hallmann, Breuer and Kühnreich (2013) suggested that hosting an event may be more important to subjective well-being than success at the event itself. Subjective well-being was measured with one item "I am happy if German athletes win many medals at Olympic Games or World Championships" with 1 = yes, and 0 = no (Hallmann, Breuer & Kühnreich, 2013, p. 230). Offering further support to these early studies, Dolan *et al.* (2019) investigated the effect that London 2012 had on the subjective well-being of residents using a quasi-experimental design. Dolan *et al.* (2019) primarily investigated subjective well-being using a single-item measure for life satisfaction (e.g., "Overall, how satisfied are you with your life these days?", p. 4), however, also asked respondents to rate their happiness, anxiety, and eudemonia (e.g., "Overall, how happy did you feel yesterday?", p. 4) using single-item measures. Responses are measured on 11-point Likert scales. The researchers found that there was an increase in subjective well-being during the event; however, again no long-term effects were detected.

While offering a worthwhile picture of subjective well-being and sport mega-events, the aforementioned studies have relied on single-item measures of happiness to determine whether the sport event impacted subjective well-being. Littlejohn *et al.* (2016) built a scale to measure holistic subjective well-being according to Diener's (2000) holistic conceptualisation. This scale was then used by Taks and colleagues (2016) to empirically investigate the happiness of non-event attendees in the context of two non-mega sport events (NMSEs). The authors found that awareness of the sport events indicated higher happiness; however, these data were collected during one-time point and, therefore, increased or decreased levels of happiness could not be determined due to the lack of baseline (pre-event) data. Filo and Coghlan (2016) also examined sport event experiences using a holistic conceptualisation of subjective well-being. Filo and Coghlan (2016) used Seligman's (2011, 2018) PERMA framework, which involves five components of well-being: positive emotion, engagement, relationships, meaning, and accomplishment. The authors found that there was evidence of each of these five PERMA domains within individuals' experiences with charity sport events. Using qualitative data, the authors found that the participatory nature of the charity sport event, as well as the charitable aspect, amongst other features of this context contributed to each PERMA domain being evident (Filo & Coghlan, 2016). The authors called for future research using PERMA study for various event contexts to determine the link between these aspects of well-being and sport events.

Further advancing understanding of sport events and subjective well-being, scholars have begun to investigate the underlying mechanisms related to the observed increase in subjective well-being during events. In their paper, Maennig and Porsche (2008) suggested that factors such as avoiding friction (e.g., avoiding traffic and noise), the weather, and event atmosphere may have played a role in the feel-good effect resulting from the 2006 FIFA World Cup. Schlegel, Pfitzner and Koenigstorfer (2017) investigated the effect of perceived atmosphere on subjective well-being both before and during the 2014 FIFA World Cup in Rio de Janeiro. The authors found that in general, subjective well-being was higher during the event than before the event. Further, they found a positive and significant relationship between perceived atmosphere and subjective well-being during the event. These findings offer support for the potential of liminality (Chalip, 2006) related to sport events for creating a space in which individuals may experience increased subjective well-being. Schlegel, Pfitzner and Koenigstorfer's (2017) study was a starting point to fill a gap regarding the potential underlying mechanisms for subjective well-being arising from sport events. To advance understanding and learn more about the potential of sport events to influence subjective well-being and offer an SROI for taxpayers, further research regarding the underlying mechanisms related to subjective well-being, such as social experiences, is warranted as well as approaching subjective well-being more broadly.

Sport mega-events and subjective well-being: the case of Rio 2016

As demonstrated earlier, results from existing research have not been consistent, and therefore we cannot conclude whether sport events improve subjective well-being, how long such outcomes may last, or through which mechanisms these outcomes are possible. Further, no research to date has investigated social impact experiences as potential mechanisms for subjective well-being. As such, some advancements can be made in the pursuit of such answers. In the context of sport events and social impacts, investigating subjective well-being in relation to event experiences offers some additional insight into the effect events may have on host country residents. To begin to fill this identified gap in the sport event and subjective well-being literature, we conducted a study during the 2016 Olympic Games, hosted by Rio de Janeiro, Brazil, in August 2016 (abbreviated as Rio 2016 hereafter). For additional contextual information regarding Rio 2016, we refer to Rocha's (2020) work that explained in detail the complex socio-economic environment within which Rio 2016 was bid for and ultimately hosted. For the sake of the discussion in this chapter, we investigated if social impact experiences from Rio 2016 could explain some variance in subjective well-being of residents, and if these effects are sustained over time, comparing measures taken during and six months after the Games.

Methods

We used two convenience samples of Brazilians, who had not attended Rio 2016 in person but reported high involvement with the Games via media. We investigated different samples during ($n = 402$) and six months after ($n = 401$) the Games. Although different, the samples have some common characteristics. Respondents were young adults ($M_{during} = 27.9$; $SD = 6.0$; $M_{after} = 29.4$; $SD = 6.1$), all respondents had at least a bachelor's degree, and only 23% of the sample, in both times, expressed that paying the bills was difficult. The number of female respondents was close (45% during; 53% after the Games). It is important to note that a limitation of this sample is that it is not representative of the Brazilian population in regard to education and economic demographics.

We investigated five different manifestations of social impact experiences: social cohesion (four items from Taks & Rocha, 2017; e.g., "I create new friendships/relationships in the community because of the Olympic Games"), community spirit (three items adapted from Gibson *et al.*, 2014; e.g., "I feel proud that the Rio is hosting the Olympic Games"), feelings of (un)safety (three items adapted from Kim *et al.*, 2015; e.g., "Rio 2016 makes me feel unsafe because of potential terrorist attacks"), community involvement (three items adapted from Peterson, Speer & McMillan, 2008; e.g., "I discussed the organisation of the Olympic

Games with other people in the community"), and sport participation (three items from Taks & Rocha, 2017; e.g., "My interest in sport and/or physical activity has increased because of the Olympic Games").

Subjective well-being was represented by the average of three indicators: happiness (one item adapted from Kavetsos & Szymanski, 2010), life satisfaction (three items from Connolly, 2013), and life affects (eight items from Diener, 2000). Happiness was measured by asking respondents: "Taking all aspects of your life into account, please select your current overall level of happiness", measured on a 6-point Likert scale (1 – not happy at all to 6 – very happy). Life satisfaction was measured by the average of three items (Connolly, 2013), where respondents were asked to rate their current level of satisfaction with three domains – life at home, health, and occupation – based on a 4-point Likert scale (1 – not at all satisfied to 4 – very satisfied). Life affects were measured by the average of three positive items (happy, friendly, enjoying myself) and five reversed negative items (frustrated, depressed, hassled, worried, tired; Diener, 2000), where respondents were asked to answer the question "How have you generally felt during the past week?", based on a 7-point Likert scale (0 – not at all to 6 – very much).

We ran two multiple regression analyses, one for data during Rio 2016 and another for data after Rio 2016 to check if social impact experiences (independent variables) could explain subjective well-being (dependent variable). We ran independent sample *t*-tests to compare social impact experiences and well-being during and six months after the Games.

Results

Regression equation for data *during* Rio 2016 was significant (F = 2.563; p = 0.27). However, the variance explained in subjective well-being was extremely small (Adj R^2 = 0.020). Moreover, none of the manifestations of social impact experiences were significant predictors of subjective well-being: social cohesion (b = −0.003; t = −0.095; p = 0.924), community spirit (b = 0.033; t = 1.145; p = 0.254), feelings of (un)safety (b = −0.040; t = −1.873; p = 0.062), community involvement (b = 0.040; t = 1.683; p = 0.093), and sport participation (b = 0.003; t = 0.140; p = 0.889).

Regression equation for data *after* Rio 2016 confirmed the result during the Games. The equation was significant (F = 2.626; p = 0.24), once more, with an extremely low percentage of variance explained in subjective well-being by the combination of all social impact experiences (Adj R^2 = 0.021). None of the manifestations of social impact experiences were significant predictors of subjective well-being: social cohesion (b = −0.029; t = −0.740; p = 0.460), community spirit (b = 0.073; t = 2.397; p = 0.017), feelings of (un)safety (b = −0.018; t = −0.893; p = 0.372), community involvement (b = 0.013; t = 0.466; p = 0.642), and sport participation (b = 0.009; t = 0.360; p = 0.719). To control for multiple comparisons, we considered an alpha error of 0.01.

Results of the *t*-tests show that all manifestations of social impact experiences had lower values six months after Rio 2016: social cohesion (ΔM = 0.870;

Table 2.1 Differences between during and after Rio 2016 in social impacts and subjective well-being

		During Rio 2016		*After Rio 2016*				
		M	*SD*	*M*	*SD*	*t*	*df*	*p*
Social impact Experiences[(1)]	Social cohesion	3.47	1.45	2.6	1.5	8.095	755	<0.001
	Community spirit	4.04	1.66	3.39	1.68	12.492	755	<0.001
	Feelings of (un)safety	3.55	1.59	3.24	1.67	2.604	755	0.009
	Community involvement	4.50	1.72	3.31	1.52	10.097	755	<0.001
	Sport participation	3.87	1.96	2.96	1.87	6.502	754	<0.001
Subjective well-being		4.15	0.64	4.13	0.65	0.343	801	0.732
	Happiness[(2)]	4.61	0.86	4.64	0.78	−0.499	801	0.618
	Life satisfaction[(3)]	2.98	0.49	2.97	0.49	0.558	801	0.577
	Affects[(4)]	4.86	0.98	4.79	1.01	0.915	798	0.361

Note: [(1)]measured on a 7-point Likert scale from 1 – "totally disagree" to 7 – "totally agree"; [(2)]measured on a 6-point Likert scale from 1 – "not happy at all" to 6 – "very happy"; [(3)]measured on a 4-point Likert scale from 1 – "not at all satisfied" to 4 – "very satisfied"; [(4)]measured on a 7-point Likert scale from 0 – "not at all" to 6 – "very much".

$t_{755} = 8.095$; $p < 0.001$), community spirit ($\Delta M = 1.437$; $t_{755} = 12.492$; $p < 0.001$), feelings of (un)safety ($\Delta M = 0.309$; $t_{755} = 2.604$; $p = 0.009$), community involvement ($\Delta M = 1.192$; $t_{755} = 10.097$; $p < 0.001$), and sport participation ($\Delta M = 0.905$; $t_{755} = 6.502$; $p < 0.001$). However, subjective well-being did not change significantly six months after the Games ($\Delta M = 0.016$; $t_{801} = 0.343$; $p = 0.732$) (Table 2.1).

Social impact experiences from Rio 2016 do not explain variances in subjective well-being either during or after Rio 2016. As expected, social impact experiences decreased after the Games. However, the values of subjective well-being did not change. These results seem to indicate that social impacts experiences promoted by hosting Rio 2016 did not change the subjective well-being of host country residents, and subjective well-being was not influenced by Rio 2016.

Discussion

The social impact results from this study are in line with results reported elsewhere, where lasting effects of social impacts from sport events were not observed (e.g., Maennig & Porsche, 2008; Kavetsos & Szymanski, 2010), or limited long-term effects were sustained following the event (Gibson *et al.*, 2014). The results regarding subjective well-being are not consistent with previous literature regarding hosting sport events and happiness, where short-term effects were observed (e.g., Kavetsos & Szymanski, 2010). Results of the current study show that subjective well-being was not affected by social experiences with the Olympic Games in the context of Rio 2016. We understand this is unique because previous studies have not investigated social impact experiences as possible underlying mechanisms of social well-being. Some studies have assumed but have not tested effects of social experiences (e.g., Gibson *et al.*, 2014). This study extends the existing research that demonstrates there is no lasting effect following a sport mega-event on subjective well-being. We found that evaluations of subjective well-being did not change from the time of the Games (where it is supposed to be at its peak; Kavetsos & Szymanski, 2010) to six months after the Games. Therefore, we have little support to affirm that hosting Olympic Games can change subjective well-being amongst non-attendee host country residents.

This study provides insight into underlying mechanisms that may be associated with subjective well-being, which is an under-investigated area in relation to subjective well-being and events. The research approach places individuals' experiences with the event at the centre, to test how experiences can be investigated as opposed to perceptions, as well as the value this adds to investigating SROI. However, the results from the study leave areas of additional research to be explored and suggest that promises regarding long-term positive outcomes from sport mega-events may need to be tempered when promoting or supporting an event.

Considerations and future directions

Underlying mechanisms

Additional research in this area is still needed, particularly regarding actual lived experiences as underlying mechanisms, and how to go about leveraging those mechanisms for positive subjective well-being outcomes, particularly if host country residents are to receive an SROI. Existing work suggests that future research should focus on individuals' actual lived experiences of social impacts (Taks & Rocha, 2017) rather than on perceptions that are more commonly used in social impact studies (e.g., Balduck, Maes & Buelens, 2011; Kim *et al.*, 2015). Indeed, these perceptions of social impacts are often based on unsubstantiated opinions, guided by media framing (Sant & Mason, 2015); thus, these measures can be biased. While perceptions, whether accurate or not, may influence subjective well-being, we advocate for more work specifically dealing with individual experiences of events as a worthy avenue for future research.

Social leveraging is the notion that sport events can be used strategically to achieve certain social objectives within the host community and country (Chalip, 2006). Social leveraging involves capitalising on liminality and communitas, which arise from events to achieve social objectives (Chalip, 2006). Some work demonstrates the positive and significant relationship between event atmosphere and subjective well-being (see Schlegel, Pfitzner & Koenigstorfer, 2017). Additional investigation into liminality and communitas, how to build an atmosphere that fosters subjective well-being, as well as how to appropriately socially leverage the event for the host community would be welcome. Findings in these areas may help events capitalise on the opportunity of positive social experiences for positive subjective well-being outcomes for both event attendees and others; however, the longevity of these effects remains questionable.

Scholars have found a positive relationship between community benefits and subjective well-being and, therefore, have suggested that ensuring community benefits is an important factor for subjective well-being (Yolal *et al.*, 2016). However, according to our findings, social impact experiences did not deliver the necessary community benefits to affect subjective well-being. This notion indicates that resident involvement in event planning and event outcome management may generate more meaningful community benefits and could be a useful future avenue of further investigation in the context of sport events and subjective well-being. This supports Maennig and Porsche's (2008) discussion regarding the importance of limiting friction with the community to maximise the feel-good effect. Further, additional resident engagement in planning initiatives may help sport mega-event organisers better consider the cultural context in which the event is being held and, therefore, help to tailor strategies to the host country and produce more appropriate outcomes (see Diener, 2000; Misener & Mason, 2006; Kavetsos & Szymanski, 2010). Thus, moving forward, researchers may want to consider addressing involvement from an event planning and management perspective as well as involvement in the event itself to investigate events' effects on subjective well-being.

Long-term effects of sport mega-events on subjective well-being

Demonstrated in the context of the case described earlier, as well as other empirical research on the topic (e.g., Kavetsos & Szymanski, 2010; Gibson *et al.*, 2014; Dolan *et al.*, 2019), it is clear that there are limitations of sport mega-events for creating long-term social impact. In fact, our study showed that a sport mega-event did not affect subjective well-being in the short or long term. This is in line with what Diener (2000) and scholars (e.g., Suh, Diener & Fujita, 1996; Mochon, Norton & Ariely, 2008) suggested, albeit not in the context of events, namely that significant life events do not have a significant effect on the overall subjective well-being and quality-of-life assessments. Diener (2000) suggested that "intense experiences are not the cornerstone of a happy life" (p. 36). This may be an explanation for the empirical evidence that currently exists regarding sport events and subjective well-being. Similarly, scholars have suggested the notion of adaptation in relation to subjective well-being. This concept refers to the tendency that people have to adapt over time to good or bad events, therefore returning to a baseline of happiness (Suh, Diener & Fujita, 1996). This phenomenon may well occur in the context sport events however has not, to our knowledge, been investigated in this way. Adaptation to intense experiences may present differently depending on how well-being is measured and conceptualised in a specific study (i.e., a single-item happiness measure compared to a holistic approach) and could therefore offer a more nuanced understanding of well-being. Further investigation of adaptation has the potential to provide additional insight into why long-term effects of subjective well-being have not been observed to date.

Scholars in psychology have indeed debated whether well-being should be considered as distinct from happiness and discuss the possibility that happiness measures reflect individuals' short-term mood as opposed to a general sense of life satisfaction (Seligman, 2010). If, in fact, happiness and well-being should be considered distinct concepts, this may be an explanation for the lack of long-term effects measured when research is conducted using a single-item happiness measure. Although here we have demonstrated the merits of a holistic approach to measuring subjective well-being (Littlejohn *et al.*, 2016), the conceptual make-up of well-being and its relationship with happiness is a larger discussion outside of the scope of this chapter. Regardless, this is worthy of further investigation in the context of sport events in the future, as the distinction may offer more information regarding the long-term viability of a positive social effect from sport events.

We do not find an effect of sport mega-events on subjective well-being, and existing research has not determined whether or not, and under what circumstances sport mega-events could contribute to improved subjective well-being long-term, hence more research is needed. We are not advocating that effects of sport mega-events on subjective well-being exist and should last forever. For now, more research is needed to investigate if these effects exist and, if they do, how long they last. Therefore, another avenue for future studies is an investigation about the lifespan of effects of sport mega-events on subjective well-being. Knowing the existence and the lifespan of effects can help sport event organisers and host city/country policymakers to find support for hosting their events. Therefore, until some lasting effects are observed that can be attributed to sport mega-events, promises of well-being outcomes need to be tempered in public policy and messaging supporting sport mega-events, as well as justification for the use of public funding to host these types of events.

Contextual and theoretical advancements

Most of the research in regard to sport events and subjective well-being has been conducted in the context of sport mega-events (for notable exceptions, see Littlejohn *et al.*, 2016; Taks *et al.*, 2016), therefore, research in other event contexts may provide additional insight. In particular, advancing research in NMSEs may provide further opportunities to capitalise on these underlying mechanisms and, therefore, offer a promising avenue for future research. Taks *et al.* (2016) investigated an NMSEs potential to affect happiness in the host community and found that non-event attendees who were aware of the event demonstrated higher levels of happiness. However, the authors stressed that causality cannot be confirmed and these affects may not be attributed directly to the event itself (Taks *et al.*, 2016). Further, scholars have highlighted the potential of festivals for improving subjective well-being. In Yolal *et al.*'s (2016) work, for example, community benefits and subjective well-being were found to be positively associated during a film festival. Here, the authors explained how certain characteristics of festivals provide an appropriate environment for these effects to manifest, many of which are shared by NMSEs. NMSEs have the potential to be better connected to the community, potentially addressing community involvement, and enhancing community development (Taks, 2013, 2016). Yolal and colleagues (2016) suggested that lower resource demands may be important in influencing subjective well-being, while Maennig and Porsche (2008) suggested that to maximise the potential feel-good factor of a sport event, it is necessary to minimise points of contention with the host community. NMSEs have lower resource demands than their sport mega-event counterparts (Agha & Taks, 2015), and may offer the potential, therefore, to address these concerns and foster the underlying mechanisms for subjective well-being to be positively affected by sport events.

Theoretically, advancing how well-being is researched and addressed in the sport event literature may help advance some of the aforementioned areas of research and lines of inquiry. Future research in the area of well-being and sport events should further consider subjective well-being as a holistic concept, as well as other forms of well-being, namely social and psychological well-being. Although these concepts are distinct and have been studied in other fields (i.e., Keyes, 1998; Chen *et al.*, 2013), there is a limited discussion in the sport event literature regarding their differences and similarities, and how various aspects of each concept may be influenced by, or manifested in, the sport event context. Recently, scholars have suggested advancing the study of sport and well-being in general and to do so in part by better differentiating between hedonic (subjective) well-being and eudemonic (psychological) well-being in future studies (Inoue, Sato & Filo, 2020), further justifying additional research in the area.

Conclusion

Despite a variety of existing research in the realm of sport, sport events, and subjective well-being, there remain areas that warrant further investigation. There remain questions regarding how sport events may influence subjective well-being, to what extent and by way of which underlying mechanisms. Studies such as the one described here which investigate actual lived experiences and explore underlying mechanisms of observed effects can begin to fill such gaps in the existing literature. We found that social impact experiences are not related to subjective well-being in the context of a sport mega-event – Rio 2016 Olympic Games. Additionally, in terms of social impact experiences, no long-term effects were identified six months after the event. These findings are aligned with existing research in the field (e.g., Maennig & Porsche, 2008; Kavetsos & Szymanski, 2010; Gibson *et al.*, 2014), signalling an absence of effects from social interactions promoted by events to affect subjective well-being.

While addressing some gaps in existing research, this study also offers promising avenues for future research in this area. Research moving forward should continue to explore sport mega-events and subjective well-being, host resident experiences, and under what conditions certain effects occur. Specifically, scholars should investigate further the features of certain events that may influence subjective well-being and how to best leverage those features for sustained impact (e.g., atmosphere; Schlegel, Pfitzner & Koenigstorfer, 2017). Further, research in the context of smaller sport events may indicate that longer-term effects are more likely to happen in other event contexts compared to sport mega-events. Finally, future research should work to expand the investigation of well-being in sport events methodologically by conceptualising well-being in different ways (hedonic versus eudemonic; Inoue, Sato & Filo, 2020) and addressing additional other components of well-being (e.g., PERMA; Seligman, 2011, 2018) to better understand well-being in the context of sport events. Ultimately, additional research in this area will help to inform event managers and policymakers regarding the role that sport events may play in contributing to residents' well-being and, therefore, may help to inform evidence-based policy decisions that ensure an SROI for sport event host residents.

Note

1 We recognise that in general terms, the Olympic Games should be referred to as the Olympic and Paralympic Games; however, the study reported here did not include the Paralympic component of the Games; hence, we refer to the Games or Olympic Games in the remainder of the text.

References

Agha, N., & Taks, M. (2015). A theoretical comparison of the economic impact of large and small events. *International Journal of Sport Finance*. 10 (3), 199–216.

Arvidson, M., Lyon, F., McKay, S., & Moro, D. (2013). Valuing the social? The nature and controversies of measuring social return on investment (SROI). *Voluntary Sector Review*. 4 (1), 3–18.

Balduck, A.-L., Maes, M., & Buelens, M. (2011). The social impact of the Tour de France: Comparisons of residents' pre- and post-event perceptions. *European Sport Management Quarterly*. 11 (2), 91–113.

Chalip, L. (2006). Towards social leverage of sport events. *Journal of Sport & Tourism*. 11 (2), 109–127.

Chen, F.F., Jing, Y., Hayes, A., & Lee, J.M. (2013). Two concepts or two approaches? A bifactor analysis of psychological and subjective well-being. *Journal of Happiness Studies*. 14 (3), 1033–1068.

Connolly, M. (2013). Some like it mild and not too wet: The influence of weather on subjective well-being. *Journal of Happiness Studies*. 14 (2), 457–473.

Davies, L.E., Taylor, P., Ramchandani, G., & Christy, E. (2019). Social return on investment (SROI) in sport: A model for measuring the value of participation in England. *International Journal of Sport Policy and Politics*. 11 (4), 585–605.

Deery, M., & Jago, L. (2010). Social impacts of events and the role of anti-social behaviour. *International Journal of Event and Festival Management*. 1 (1), 8–28.

Diener, E. (2000). Subjective well-being: The science of happiness and a proposal for a national index. *American Psychologist*. 55 (1), 34–43.

Diener, E., & Seligman, M.E.P. (2004). Beyond money: Toward an economy of well-being. *Psychological Science in the Public Interest*. 5 (1), 1–31.

Dolan, P., Kavetsos, G., Krekel, C., Mavridis, D., Metcalfe, R., Senik, C., Szymanski, S., & Ziebarth, N.R. (2019). Quantifying the intangible impact of the Olympics using subjective well-being data. *Journal of Public Economics*. 177, 1–14.

Dolan, P., Peasgood, T., & White, M. (2008). Do we really know what makes us happy? A review of the economic literature on the factors associated with subjective well-being. *Journal of Economic Psychology*. 29 (1), 94–122.

Downward, P., Hallmann, K., & Rasciute, S. (2018). Exploring the interrelationship between sport, health and social outcomes in the UK: Implications for health policy. *European Journal of Public Health*. 28 (1), 99–104.

Downward, P., & Rasciute, S. (2011). Does sport make you happy? An analysis of the well-being derived from sports participation. *International Review of Applied Economics*. 25 (3), 331–348.

Filo, K., & Coghlan, A. (2016). Exploring the positive psychology domains of well-being activated through charity sport event experiences. *Event Management*. 20 (2), 181–199.

Getz, D. (2012). Event studies: Discourses and future directions. *Event Management*. 16 (2), 171–187.

Getz, D., & Page, S.J. (2016). *Event studies: Theory, research, and policy for planned events* (3rd ed.). Oxon, UK: Routledge.

Gibson, H.J., Walker, M., Thapa, B., Kaplanidou, K., Geldenhuys, S., & Coetzee, W. (2014). Psychic income and social capital among host nation residents: A pre–post analysis of the 2010 FIFA world cup in South Africa. *Tourism Management*. 44, 113–122.

Grix, J., & Carmichael, F. (2012). Why do governments invest in elite sport? A polemic. *International Journal of Sport Policy and Politics*. 4 (1), 73–90.

Hallmann, K., Breuer, C., & Kühnreich, B. (2013). Happiness, pride and elite sporting success: What population segments gain most from national athletic achievements? *Sport Management Review*. 16 (2), 226–235.

Heere, B., Walker, M., Gibson, H., Thapa, B., Geldenhuys, S., & Coetzee, W. (2013). The power of sport to unite a nation: The social value of the 2010 FIFA World Cup in South Africa. *European Sport Management Quarterly*. 13 (4), 450–471.

Inoue, Y., Sato, M., & Filo, K. (2020). Transformative sport service research: Linking sport services with well-being. *Journal of Sport Management*. 34 (4), 1–6.

Karadakis, K., & Kaplanidou, K. (2012). Legacy perceptions among host and non-host Olympic Games residents: A longitudinal study of the 2010 Vancouver Olympic games. *European Sport Management Quarterly*. 12 (3), 243–264.

Kavetsos, G., & Szymanski, S. (2010). National well-being and international sports events. *Journal of Economic Psychology*. 31 (2), 158–171.

Késenne, S. (2005). Do we need an economic impact study or a cost-benefit analysis of a sports event? *European Sport Management Quarterly*. 5 (2), 133–142.

Késenne, S. (2012). The economic impact, costs and benefits of the FIFA World Cup and the Olympic Games: Who wins, who loses? In: Maennig, W., & Zimbalist, A. (eds.) *International handbook on the economics of mega sporting events*. Cheltenham, UK: Edward Elgar Publishing, pp. 270–278.

Keyes, C.L.M. (1998). Social well-being. *Social Psychology Quarterly*. 61 (2), 121–140.

Kim, J., & James, J.D. (2019). Sport and happiness: Understanding the relations among sport consumption activities, long- and short-term subjective well-being, and psychological need fulfillment. *Journal of Sport Management*. 33 (2), 119–132.

Kim, W., Jun, H.M., Walker, M., & Drane, D. (2015). Evaluating the perceived social impacts of hosting large-scale sport tourism events: Scale development and validation. *Tourism Management*. 48, 21–32.

Littlejohn, M., Taks, M., Wood, L., & Snelgrove, R. (2016). Sport events and happiness: Towards the development of a measuring instrument. *World Leisure Journal*. 58 (4), 255–266.

Mackintosh, C., Darko, N., & May-Wilkins, H. (2016). Unintended outcomes of the London 2012 Olympic games: Local voices of resistance and the challenge for sport participation leverage in England. *Leisure Studies*. 35 (4), 454–469.

Maennig, W., & Porsche, M. (2008). The feel-good effect at mega sport event: Recommendations for public and private administration informed by the experience of the FIFA World Cup 2006. *Hamburg Contemporary Economic Discussion Paper No. 18*, 1–27.

McCloy, C. (2009). Canada hosts the world: An examination of the first federal sport hosting policy (1967). *The International Journal of the History of Sport*. 26 (9), 1155–1170.

McCloy, C., & Thibault, L. (2013). Hosting policies of sport events. In: Thibault, L., & Harvey, J. (eds.) *Sport policy in Canada*. Ottawa, ON: University of Ottawa Press, pp. 243–265.

Mochon, D., Norton, M.I., & Ariely, D. (2008). Getting off the hedonic treadmill, one step at a time: The impact of regular religious practice and exercise on well-being. *Journal of Economic Psychology*. 29 (5), 632–642.

Misener, L., & Mason, D.S. (2006). Creating community networks: Can sporting events offer meaningful sources of social capital? *Managing Leisure*. 11 (1), 39–56.

Müller, M. (2015). The mega-event syndrome: Why so much goes wrong in mega-event planning and what to do about it. *Journal of the American Planning Association*. 81 (1), 6–17.

Nicholls, J., Lawlor, E., Neitzert, E., & Goodspeed, T. (2012). *The guide to social return on investment*. London: The SROI Network: Accounting for Value. Available from www.socialvalueuk.org/app/uploads/2016/03/The%20Guide%20to%20Social%20Return%20on%20Investment%202015.pdf [Accessed 10 June 2020].

Panza, G.A., Taylor, B.A., Thompson, P.D., White, C.M., & Pescatello, L.S. (2017). Physical activity intensity and subjective well-being in healthy adults. *Journal of Health Psychology*. 24 (9), 1257–1267.

Pawlowski, T., Downward, P., & Rasciute, S. (2014). Does national pride from international sporting success contribute to well-being? An international investigation. *Sport Management Review*. 17 (2), 121–132.

Peterson, N.A., Speer, P.W., & McMillan, D.W. (2008). Validation of A brief sense of community scale: Confirmation of the principal theory of sense of community. *Journal of Community Psychology*. 36 (1), 61–73.

Preuss, H., Andreff, W., & Weitzmann, M. (2019). *Cost and revenue overruns of the Olympic Games 2000–2018*. Wiesbaden, DE: Springer Gabler.

Rasciute, S., & Downward, P. (2010). Health or happiness? What is the impact of physical activity on the individual? *Kyklos*. 63 (2), 256–270.

Rocha, C.M. (2020). Temporal variations in the relationship between legacies and support: A longitudinal case study in Rio 2016 Olympic Games. *Journal of Sport Management*. 34 (2), 130–146.

Sant, S.-L., & Mason, D.S. (2015). Framing event legacy in a prospective host city: Managing Vancouver's Olympic bid. *Journal of Sport Management*. 29 (1), 42–56.

Sato, M., Jordan, J.S., & Funk, D.C. (2015). Distance running events and life satisfaction: A longitudinal study. *Journal of Sport Management*. 29 (4), 347–361.

Schlegel, A., Pfitzner, R., & Koenigstorfer, J. (2017). The impact of atmosphere in the city on subjective well-being of Rio de Janeiro residents during (vs. before) the 2014 FIFA world cup. *Journal of Sport Management*. 31 (6), 605–619.

Seligman, M. (2010). Flourish: Positive psychology and positive interventions. In: *The tanner lectures on human values*. Ann Arbour, MI, The University of Michigan, pp. 231–243.

Seligman, M. (2011). *Flourish*. New York: Free Press.

Seligman, M. (2018). PERMA and the building blocks of well-being. *The Journal of Positive Psychology*. 13 (4), 333–335.

Suh, E., Diener, E., & Fujita, F. (1996). Events and subjective well-being: Only recent events matter. *Journal of Personality and Social Psychology*. 70 (5), 1091–1102.

Taks, M. (2013). Social sustainability of non-mega sport events in a global world. *European Journal for Sport and Society*. 10 (2), 121–141.

Taks, M. (2016). The rise and fall of mega sport events: The future is in non-mega sport events. In: Vanden Auweele, Y., Cook, E., & Parry, J. (eds.) *Ethics and governance in sport: The future of sport imagined*. New York: Routledge, pp. 84–93.

Taks, M., Littlejohn, M., Snelgrove, R., & Wood, L. (2016). Sport events and residential happiness: The case of two non-mega sport events. *Journal of Global Sport Management*. 1 (3–4), 90–109.

Taks, M., & Rocha, C. (2017). Maakte Rio 2016 de Brazilianen gelukkig? [Did Rio 2016 make the Brasilians happy?] In: Hover, P., & Breedveld, K. (eds.) *The story of Rio 2016: De maatschappelijke betekenis van de Olympische en Paralympische spelen 2016 [The story of Rio 2016: The societal meaning of the Olympic and Paralympic games 2016]*. Utrecht, Nl: Mulier Instituut, pp. 119–127.

Weed, M., Coren, E., Fiore, J., Wellard, I., Chatziefstathiou, D., Mansfield, L., & Dowse, S. (2015). The Olympic games and raising sport participation: A systematic review of evidence and an interrogation of policy for a demonstration effect. *European Sport Management Quarterly*. 15 (2), 195–226.

Yolal, M., Gursoy, D., Uysal, M., Kim, H. (Lina), & Karacaoğlu, S. (2016). Impacts of festivals and events on residents' well-being. *Annals of Tourism Research*. 61, 1–18.

3 Effectively utilising professional sport franchise home-games as platforms for enhancing peri- and post-event well-being outcomes

Rachel J. Batty

Introduction

Much of the research on sport and well-being has, to date, focused on athletes and sport event participants (Steptoe & Butler, 1996; Downward & Rascuite, 2011; Dunn, 2014; Zhou & Kaplanidou, 2018). However, emerging research is suggesting that attendance at major sports events also has a positive association with spectator well-being, especially in regard to life satisfaction and happiness (Ramchandani, Coleman & Millar, 2019). This chapter adds to emerging discourse by exploring the well-being outcomes associated with attendance at professional sport franchise home-games. The investigation also considers the generation of well-being outcomes for spectators, peri (i.e. during) event and post-event. This chapter begins with a literature review that offers insight into existing research and perceptions on sport and well-being, the well-being outcomes associated with sport events and how such outcomes can be enhanced through strategic leveraging and event legacy planning. A case study of the Auckland Tuatara professional baseball franchise and the associated National Sporting Body (NSB), Baseball New Zealand, is utilised to provide context to the discussion.

Literature review

This literature review is framed around the three key themes pertaining to the noted case study. Those three themes are well-being, leveraging and event legacy. These themes are defined and positioned within a sporting context and explored in relation to their interconnectedness within a sport event setting.

Well-being and sport

A positive relationship between sport and well-being has long been widely accepted (Mayoh & Jones, 2015) to the point where sport is widely recognised as contributing to the betterment of society (Duarte, Alonso & O'Shea, 2012; Warner & Dixon, 2016). Such betterment is frequently linked to a range of social outcomes, with an emphasis on well-being. Such outcomes include the overall happiness, quality of life, fulfilment, positive relations and acceptance by others

(Downward & Rascuite, 2011; Lundqvist, 2011; Lee, Cornwell & Babiak, 2012). Motivation and the need for relatedness often manifest in sport through access to and the showcasing of role models and can also contribute to general well-being (Grabs *et al.*, 2016).

Much of the literature pertaining to well-being as an outcome of sport acknowledges the notion of subjective well-being (Brown *et al.*, 2010; Lundqvist, 2011; Mayoh & Jones, 2015; Ramchandani, Coleman & Millar, 2019). Subjective well-being (SWB) applies to the individual standards on which a person evaluates his or her perceived well-being (Lundqvist, 2011) and the varied emotional responses to specific life domains (Brajša-Žganec, Merkaš & Šverko, 2011). Brown *et al.* (2010) refer to SWB as the personally constructed meanings attributed to particular experiences. Ramchandani, Coleman and Millar (2019) suggest that SWB evaluations can be based on three viewpoints. Evaluative viewpoints tend to focus on an appraisal or reflection of a situation. Experiential viewpoints refer to positive and negative emotional experiences, over a short time frame. Eudemonic viewpoints associate a sense of meaning or purpose to an experience. Sport and SWB can be experienced in two primary ways: via sport participation and sport spectatorship (Kim & James, 2019).

Research on sport participation acknowledges well-being outcomes that predominantly relate to physical and mental (or psychological) health benefits (Khan *et al.*, 2012; Downward & Dawson, 2016; Edwards & Rowe, 2019). The World Health Organization (2003) specifically states that sport can help reduce the risk of heart disease and diabetes, control weight, build healthy bones, muscles and joints, promote psychological well-being and prevent and control risky behaviours, especially amongst children and young people. Any ability for sport to promote positive health outcomes is somewhat dependent on the community setting and an individual's circumstances, including demographics and psychographics (Steptoe & Butler, 1996; Downward & Rascuite, 2011; Zhou & Kaplanidou, 2018; Edwards & Rowe, 2019). Research within this area consequently tends to place emphasis on athlete well-being (Dunn, 2014).

Research pertaining to well-being and sport spectatorship has begun to emerge more recently. Well-being outcomes in this context are primarily noted as psychological. Such outcomes tend to revolve around social connection, positive emotions, life satisfaction, motivation and happiness (Kavetsos & Szymanski, 2010; Brajša-Žganec, Merkaš & Šverko, 2011; Marks *et al.*, 2011; Grabs *et al.*, 2016; Spaaij, Oxford & Jeanes, 2016). Extended research within this area acknowledges the three SWB viewpoints noted by Ramchandani, Coleman and Millar (2019). For example, sport event attendance can lead to the creation of memorable experiences that can then be reflected upon (Spracken & Lamond, 2016), thus emphasising the evaluative nature of SWB (Ramchandani, Coleman & Millar, 2019). Attending live sports games can help sports fans better understand themselves and bring meaning and purpose to their lives (Fernandez & Walker, 2019), thus emphasising the eudemonic nature of SWB (Ramchandani, Coleman & Millar, 2019). Attending a sports game involving a local team, as a form of social activity, can increase perceptions of emotional support (Inoue *et al.*, 2020), thus emphasising the experiential nature of SWB (Ramchandani, Coleman & Millar, 2019).

In taking a closer look at the well-being outcomes associate with sport events, Pawlowski, Downward and Rascuite (2014) posit that while attending sport events can generate feelings of happiness, more often than not, it is the action of attending as opposed to the team success that has the bigger impact. Lee, Cornwell and Babiak (2012) suggest that, when effectively designed, a sport event can heighten that sport's impact on society. Thus, sport events can directly foster well-being through collective identity, a sense of belonging and the development of positive community attitudes (Djaballah, Hautbois & Desbordes, 2015; Spaaij, Oxford & Jeanes, 2016; Kim & James, 2019). This concept of togetherness can also be associated with aspects of social cohesion. For example, Schulenkorf (2012) states that by engaging with sport and sport events, disadvantaged or divided communities are provided with an opportunity to help themselves through newly established connections.

The pathway between sport spectatorship and sport participation has been examined, with confirmation that sport spectatorship has a positive impact on both a sense of belonging and increased consideration of sport participation (Kim & James, 2019). Sport events inspire spectators to emulate the athletes they see, at least to the extent of engaging in the sport (Khan *et al.* 2012; Mayoh & Jones, 2015; Chalip *et al.*, 2017). Research findings presented by Ramchandani, Coleman and Millar (2019, p. 13) further support this notion by confirming that

> [e]vents can serve as a platform for inspiring audiences to increase their physical activity levels, which appears to have a positive knock-on effect on their well-being. . . . Providing spectators with an inspirational experience at events might involve giving them the opportunity to meet elite athletes and hear about how they got started in sport, organising taster sessions, and signposting spectators to opportunities to participate in sport.

The conclusions of Lee, Cornwell and Babiak (2012) are of important note here, as they conclude that there are stronger links between increased sport participation and sport event attendance as opposed to spectators simply watching a sports event on television.

Sport event leveraging

Inoue *et al.* (2020) suggest that well-being outcomes can be enhanced through leveraging the potential of spectator sport events. Leveraging activities can assist in capturing and communicating the social impact of sports to the general public and society at large (Lee, Cornwell & Babiak, 2012, p. 38). Leveraging is considered as a process of strategic planning to ensure the maximisation of both short-term and long-term event outcomes (O'Brien & Chalip, 2007). In this sense,

> [l]everaging treats an event as an addition to the product and service mix at the host destination. What an event enables is, therefore, a consequence of

the ways that it is incorporated into that mix and used strategically to reach whatever strategic objectives have been set.

(Taks *et al.*, 2017, p. 186)

While sport events are frequently linked to positive economic, psychological, socio-cultural and political impacts (Schulenkorf & Edwards, 2012), there is often an assumption that such outcomes will automatically occur (Chalip *et al.*, 2017). However, event leveraging needs to be tailored to the setting in which it is applied and incorporate the pre-identification of potential shared benefits (Khan *et al.*, 2012; Schulenkorf & Edwards, 2012; Djaballah, Hautbois & Desbordes, 2015; Chalip *et al.*, 2017; Zhou & Kaplanidou, 2018). Sport organisations, therefore, need to take the initiative and responsibility for amplifying well-being outcomes rather than exclusively focusing on competition (Edwards & Rowe, 2019). Leveraging can then play an integral role in ensuring well-being outcomes are achieved (Schulenkorf, 2012; Edwards, 2015; Bell & Daniels, 2019; Edwards & Rowe, 2019). Despite this identified opportunity, it has been noted that, more often than not, sport organisations take little advantage of sport events to build participation. 'It is only after the event is over that the sport organisations recognise that they have missed an opportunity' (Taks *et al.*, 2017, p. 196). Researchers have conceded that more qualitative and quantitative research is needed on the leveraging of smaller scale sport events to better establish community-related well-being outcomes (Duarte, Alonso & O'Shea, 2012; Rowe *et al.*, 2013; Djaballah, Hautbois & Desbordes, 2015; Mayoh & Jones, 2015).

In the case of sport events that involve professional sports teams, leveraging efforts can build fan loyalty, expand fan bases, align with broader social marketing aims and establish connections with their extended community (Yancey *et al.*, 2009; Warner & Dixon, 2016). For sport organisations, utilising the connection between a team, their home base and the community can go a long way to fulfilling strategic goals and objectives (Yancey *et al.*, 2009; Robson & France, 2011; Bester, 2012; Djaballah, Hautbois & Desbordes, 2015). For this reason, sport organisations have increasingly attempted to utilise event projects for community development purposes (Schulenkorf, 2012). Consequently, effective event leveraging also relies on effective partnerships. Partnerships enable communities and corporations to pool resources to address strategies and goals common to their consumer base (Yancey *et al.*, 2009). A strategic approach to event leveraging needs to acknowledge and work with the various stakeholders associated with the sport (Zhou & Kaplanidou, 2018). Schulenkorf (2012) argues that representative groups of a sport, including the wider community, need to strategically cooperate to effectively leverage any well-being impacts. For example, sports organisations and the athletes associated with professional franchises are natural allies. 'It is a win-win situation for franchises seeking to be socially responsible and . . . those seeking to leverage resources to carry out their mission' (Yancey *et al.*, 2009, p. 324). Furthermore, professional sports have an advantage, in that they can leverage from the popularity, and profile of the sport, and their brand (Robson & France, 2011).

> The challenge, then, is to identify the best means to obtain desired public value from teams. There needs to be a shift of focus from impact to leverage. The level and nature of a team's impact will depend on the strategies and tactics that have been employed to leverage the team, not on the mere fact that a team is (or is not) present in the community.
>
> (Sparvero & Chalip, 2007, p. 23)

In situations where a particular sport may wish to optimise a sports team or franchise to enhance engagement in and awareness of their sport, they may need to 'piggyback' on professional sport and develop effective relationships in order for their situation to improve (Robson & France, 2011). Heere and James (2007), Thomson, Schlenker and Schulenkorf (2013) and Ramchandani, Coleman and Millar (2019) argue that the examination of relationships between sports teams and their surrounding community remains largely unexplored. Further research is needed on the strategic leveraging of sport events so as to maximise associated community and well-being outcomes (Thomson, Schlenker & Schulenkorf, 2013). Efforts to study the notion of community and sports teams have tended to focus, more so, on the bond between a fan and a specific team (Wann *et al.*, 2001). The case study presented in this chapter aims to address this gap in the literature.

Sport event legacies

As with many approaches to leveraging, event legacy is linked with the longer-term focus and, impacts of an event (Thomson, Schlenker & Schulenkorf, 2013). For example, Zhou and Kaplanidou (2018, p. 501) argue that leveraging strategies should be deployed to generate and sustain the social outcomes derived from the sport events in the wider community. Such wider community impacts and sustained social prosperity are also frequently referenced as elements of sporting event legacy (Misener & Mason, 2006). Event legacies tend to be frequently aligned with well-being outcomes, including community health, the betterment of society and enhanced cultural interaction (Duarte, Alonso & O'Shea, 2012; Warner & Dixon, 2016; Thomson *et al.*, 2019; Edwards & Rowe, 2019). This is because sport events have the potential to substantially impact different groups within the community (Misener & Mason, 2006).

Establishing an event legacy also requires pre-planning, with goals established from the earliest stages of planning so that any vision can be achieved (Thomson, Schlenker & Schulenkorf, 2013). The planning phase also needs to strengthen ties between community stakeholders, whilst focusing on 'shared community outcomes, accessible legacy programmes, and social inclusion' (Misener & Mason, 2006, p. 46). In this sense, sport events should 'fit' with any national sporting strategies (Yancey *et al.*, 2009; Bester, 2012; Lee, Cornwell & Babiak, 2012). To reiterate the position of Taks *et al.* (2017, p. 196), 'cooperation between the event organisers and those who are leveraging is clearly essential in order for goals to be shared and tactical implementations to be enabled'.

While increasing attention has been placed on the legacies aligned with large-scale sport events, such as the Commonwealth and Olympic Games (Preuss, 2010; Cornelissen, Bob & Swart, 2011; Kassens-Noor *et al.*, 2015; Thomson *et al.*, 2019), Djaballah, Hautbois and Desbordes (2015) suggest that smaller, non-mega, events may provide improved and maintainable social outcomes for local communities. In adding to this notion, Sparvero and Chalip (2007) confirm that research in this area tends to focus on economic impact and needs to incorporate a broader perspective that considers social objectives. Such ongoing research will further support the view that sport is one sector that can improve the health of a nation (Khan *et al.*, 2012).

Further addressing the gaps in existing literature

In addition to the calls for further research in the areas noted earlier, there has also been limited research on the sport of baseball and its links between tournaments, well-being and legacy. Research from the late 1990s suggested that 'minor league teams and their stadiums can be contributors to a community's economic well-being and quality of life' (Johnson, 1998, p. 50). Klein's (2009) research focused more directly on the social benefits of sport in the context of Baseball in the Dominican Republic and acknowledged aspects of player education and sport for development initiatives. More recently, Kihl, Babiak and Tainsky's (2014) evaluation of a Major League Baseball team in Midwest USA opted to focus on the evaluation of corporate community initiatives and issues with quality implementation, yet highlighted 'the need to critically examine the delivery of youth sports programs in general and in particular youth programs delivered in partnership with professional sport franchises' (Kihl, Babiak & Tainsky, 2014, p. 335). Rhodes (2020) reviewed corporate social responsibility decision-making associated with Major League Baseball teams, however, her research focused on business and charitable notices. To date, there has been no research on the leveraging opportunities and well-being outcomes associated with a professional baseball franchise in the context of a New Zealand sport setting.

This chapter aims to contribute to discourse on sport events well-being outcomes and attempts to address some of the noted existing gaps in the literature. It presents findings on the perceived well-being outcomes from a home-game event series, as part of a wider-serving case study on the Auckland Tuatara and Baseball New Zealand.

Case study overview

The Auckland Tuatara baseball franchise was established in 2018 as New Zealand's first professional baseball team. The team's emergence came about following an invitation from the Australian Baseball League (ABL) for countries within the Pacific region to join their expanding league ('Baseball: Auckland Tuatara

Officially Unveiled', 2018). During their inaugural season, the Auckland Tuatara played the majority of their games in Australia and only hosted a small handful of games at a temporary venue in Auckland (New Zealand). This resulted in limited opportunities for the franchise to fully engage with its fans and sustain strong links with the local community. Ahead of the 2019/2020 second season, the franchise moved into a permanent home at Auckland's North Harbour Stadium. The stadium offered the franchise with a chance to host a wider range of home games across the length of the baseball season, that is five home-game series, with each series made up of four games across four days ('Auckland Tuatara looking for better roster', 2019). The set-up on the main field, within the stadium, included the baseball diamond and a separate sports activity zone for young children, that offered a range of baseball activities (e.g. throwing, catching, running and batting). Additional home games followed the initially scheduled five home-game series when the team successfully secured its place in the playoffs ('Baseball: Auckland Tuatara qualify for ABL playoffs', 2020).

With the full support of the NSB, Baseball New Zealand, the Auckland Tuatara was considered as a means to promote participation in and awareness of the sport of baseball, inspire young players and engage and grow a community of baseball supporters. The franchise and the NSB are currently linked via a shareholding relationship that calls for assimilated values and outcomes. Shared well-being outcomes (based on the NSB Strategic Plan) include supporting youth participation in sport; promoting the benefits and values (such as health and personal development) associated with baseball participation; and enhancing community-based partnerships (Baseball New Zealand, 2019).

The following section outlines the methodological approach utilised for the case, including the study project design, data collection and data analysis phases.

Methodology

A number of considerations led to the choice and design of the methodology for this research.

For example, Coghlan and Filo (2013) specifically acknowledge the applicability of a qualitative research approach within the context of sport events and the need to understand event experiences and impacts. Stake (1995) and Gratton and Jones (2010) support the use of case studies within a sport setting, especially when it is useful to illustrate events why and how events take place within a particular setting. In acknowledging the varying viewpoints of SWB, Kavetsos and Szymanski (2010) suggest that surveys that attempt to gather direct data on an individual's perception of their own happiness or sense of belonging can be misleading. The authors go on to suggest that it can often be more appropriate for surveys to establish broader reflection on the expectations of well-being or well-being impacts on other individuals or groups. Consequently, a qualitative case study was utilised to measure (a) the perceived well-being impacts that the Auckland Tuatara home-game events were having on event attendees, peri-event and (b) the potential for extended well-being impacts post-event. The aforementioned considerations are reflected through the data collection and analysis stages of the research (and detailed later).

Data collection method

Surveys were identified as the most appropriate data collection tool in relation to collating the perspectives of home-game attendees and for the underlying purpose of the research. In acknowledging the work of Silverman (2004) and Bell (2005), surveys were deemed a reliable and quick method to collect information from multiple respondents in an efficient and timely manner, especially where time was a noted constraint (in terms of both the length of the game and the time it took for respondents to complete the survey). Brandl-Bredenbeck and Kampfe (2012) acknowledge that surveys meet the requirement of representativeness.

Surveys took the form of a hard-copy one-page document, presented on clipboards with pens provided. This method is noted by Brandl-Bredenbeck and Kampfe (2012) as still applicable and favourable, despite advances in data collection technology. The chosen method ensured that surveys were completed while respondents were attending the game. Any technical issues that may have arisen through the use of a mobile phone application were avoided (e.g. log-in errors or screen viewing in the direct sun). Such an approach also supports Gournelos, Hammonds and Wilson (2019) notion that sampling methods and survey types go hand in hand.

Survey content

The survey comprised 12 mixed qualitative and quantitative short answer/check box questions. Closed questions provided predetermined answers to which respondents could select from, using checkboxes. 'The major advantage of these [types of] questions are that they are easy for respondents to answer and easy for researchers to analyse statistically and replicate' (Brandl-Bredenbeck & Kampfe, 2012, p. 180). The inclusion of a number of open questions gave participants the opportunity to explain about their attitudes and experiences and raise new points and issues (Brandl-Bredenbeck & Kampfe, 2012; Gournelos, Hammonds & Wilson, 2019).

Survey questions with a well-being focus, included a quantitative check-box question, as well requests for open-ended qualitative feedback. Firstly, respondents were asked to identify (by way of checking a box) well-being outcomes they deemed applicable to the event (if any). The check-box options respondents could select from represented ideal outcomes associated with the event and included: Encouraging Participation; Enhancing Baseball Knowledge/Values; Creating Social Connections; Enhancing Cultural Identity and Diversity; and 'Other'. These options were selected on the basis of the identification of ideal event outcomes. Respondents were then asked to provide feedback on the perceived community benefits and outcomes they believed the home-game events and the establishment of the Auckland Tuatara had created. Lastly, respondents were asked to comment on any aspects of the event that the Auckland Tuatara and Baseball New Zealand (as the NSB) could have worked on more effectively to further enhance the well-being outcomes.

Data set

Survey respondents were randomly approached by volunteers within the stadium and asked to complete the one-page survey. Survey participation was voluntary. Survey respondents could only complete one survey across the course of the four game series. As a participation incentive, those respondents who completed all survey questions were eligible to enter a draw to win an Auckland Tuatara hooded top. Such purposive sampling aligned with the guidelines presented by Gournelos, Hammonds and Wilson (2019, p. 122) in that such an approach is suitable 'when the researcher strategically decides which people will give meaningful responses' and establishes how people feel about a specific product or service over a set time period.

Data collection

Data collection took place at the final home series of the 2019/2020 season (Brisbane Bandits versus Auckland Tuatara) at North Harbour Stadium. Surveys were collected across all four games to ensure a balanced mix of respondents and collate varying perspectives across different days and times (e.g. weeknights and weekends). Games took place consecutively on a Thursday at 7 pm, Friday at 7 pm, Saturday at 7 pm and Sunday at 1 pm.

A total of 200 survey respondents were targeted. Of those 153 surveys (76%) were completed in full and without errors. Errors included (but were not limited to) the misreading of questions and provision of incomplete responses, or unidentifiable responses.

Data analysis

Based on the set of questions specifically relating to the topic of well-being, responses were largely qualitative in nature. A such, this allowed for the identification of previously unknown explanations (Klassen *et al.*, 2012). The open-ended questions allowed for the sourcing of rich description, thus adding to the validity of the primarily qualitative approach (Creswell & Miller, 2000). Analysis of the open-ended question responses incorporated a qualitative coding process involving identification of the frequency or variety of the most salient themes (Cohen, Manion & Morrison, 2007) and the placing of the raw data into logical meaningful categories (Edwards & Skinner, 2009). In relation to the one question that incorporated check-box responses, totals and averages were calculated and measured against accompanying qualitative responses to check for consistent understanding of the well-being topic. Such an approach is considered valid by the likes of Bryman and Bell (2011), for the purpose of congruency, but is not deemed as a mixed-method research approach in this particular case.

Case study findings

Survey respondents were initially asked to confirm if they considered themselves a pre-existing member of the New Zealand baseball community, and if so, in what capacity (e.g. player, club or regional association, inaugural season member or sponsor). This question was asked to establish any familiarity with the sport, team and Auckland Tuatara events that may have influenced differences in event outcome opinions for the home-series in question. Based on the 153 fully completed surveys, those who attended the home-game series incorporated a mix of people who already considered themselves part of the New Zealand baseball community ($N = 69$) and people who didn't consider themselves part of the New Zealand baseball community ($N = 84$).

As an introduction to the concept of well-being, respondents were asked to consider how the sport event contributed to general well-being outcomes. A series of five check-box answers was offered as responses to this initial question, whereby respondents could select as many answers (or as few) as they considered to apply. Check-box answers incorporated psychological well-being outcomes that focused on the social connection and cultural identity/diversity identified peri-event. Physical well-being outcomes referred to the encouragement of participation in the sport, post-event. The enhancement of baseball values (both those general to the sport and those specific to Baseball New Zealand) emphasised additional psychological well-being outcomes peri- and post-event. Such values, as noted in the case study overview and included inspiring young players, engaging and growing a community, and promoting the value of sport participation and are identified as congruent to well-being outcomes identified by Khan *et al.* (2012), Downward & Dawson (2016), Spaaij, Oxford & Jeanes (2016) and Kim & James (2019).

Respondent data noted in Figure 3.1 present combined totals across all four games in the home-game series. As a post-event physical well-being outcome, 'Encouraging Participation' was perceived as the most highly achieved outcome, with those who identified themselves as non-baseball community members ranking this outcome higher (by $N = 21$) than their counterparts. 'Encouraging Cultural

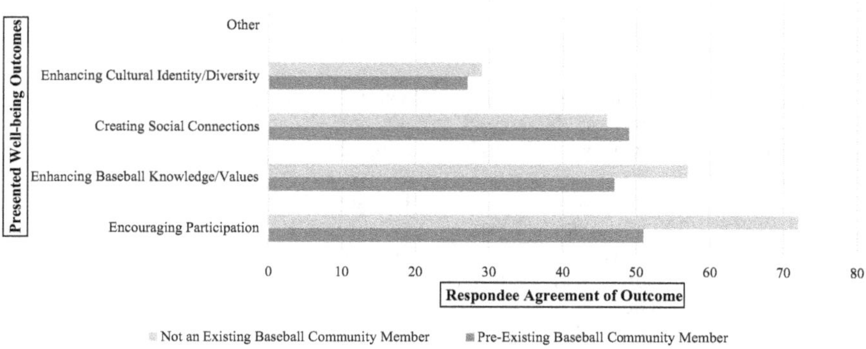

Figure 3.1 Event attendee identification of pre-specified well-being outcomes

Identity/Diversity' was perceived as the least achieved outcome, with respondents identifying as baseball community members and non-baseball community members ranking this outcome similarly. A fifth check-box option of 'other' was available for respondents to note any additional well-being outcomes if the so desired, however, no respondents selected this option.

Respondents were then asked to comment further on any community benefits or outcomes they believed the home-game events and the establishment of the Auckland Tuatara had created. These qualitative responses were grouped under four codes (based on wording and associated impact) and also classified based on their association with physical and psychological well-being (based on supporting literature) and the consideration of peri- and post-event impact. Coded findings and accompanying example quotes are presented in Table 3.1.

Worthy of note in the initial presentation of these insights is that although the previous survey question had asked respondents to identify any additional well-being outcomes, this broader request for additional event outcomes resulted in the unknown acknowledgement of additional physical and psychological well-being outcomes. For example, the event presented an opportunity to showcase role models for the sport of baseball peri-event. This was by way of the professional players providing inspiration and goal realisation to existing younger players and those

Table 3.1 Extended community benefits/outcomes associated with home-game events

Code	*Example Quotes*	*Well-Being Focus*	*Pre-/Peri-/Post-Event*
Showcases role models	'This sets an achievable goal for our young players. Attending the game allows them to see and touch the players and realise that goal'. 'Watching the game live and meeting the players at the end inspires young children'.	Psychological (Khan *et al.*, 2012; Mayoh & Jones, 2015; Grabs *et al.*, 2016; Chalip *et al.*, 2017).	Peri- and post-event
Encourages general physical activity	'Promotion of physical health and well-being through sport'. 'Shows the advantages of being active and pursuing dreams in sport'. 'The kids' area is a great idea! It gets them active and introduces basic ball skills to them'.	Physical/leading to physical (Khan *et al.*, 2012; Downward & Dawson, 2016; Edwards & Rowe, 2019; Kim & James, 2019).	Post-event

Code	*Example Quotes*	*Well-Being Focus*	*Pre-/Peri-/Post-Event*
Promotes participation and pathways in baseball	'Entices kids to become involved in baseball at a grassroots level'. 'The game is promoting the pathways available to kids playing baseball in New Zealand. Junior players can see top-level baseball close to home'. 'This event demonstrates that baseball is growing and offers pathways and opportunities across a range of socio-economic groups and provides a local opportunity to play professional baseball'.	Leading to physical (Johnson, 1998; Kim & James, 2019).	Post-event
Builds a community and promotes acceptance and diversity	'Baseball events, like this, present opportunities to bring a range of socio-economic groups together to further extend our existing baseball community'. 'There's a strong feeling of "togetherness" in the stadium'. 'This promotes the sport to all ages, cultures, genders and socio-economic groups.' 'Shows group togetherness and the diversity of New Zealanders'.	Psychological (Schulenkorf, 2012; Djaballah, Hautbois & Desbordes, 2015; Spaaij, Oxford & Jeanes, 2016; Kim & James, 2019).	Peri- and post-event

children wanting to become involved in baseball post-event. The provision of the children's activity area, to the left of the main diamond, presented children with an opportunity to actively participate in a range of sporting activities peri-event. Respondents also acknowledged the existence and development of community cohesion and the concept of 'togetherness' peri- and post-event. Lastly, sport participation pathways were referenced by respondents, in terms of supporting and enabling sustainable physical activity, promoting goal attainment and confidence through the establishment of a career in baseball and encouraging event and sport access to individuals from a range of socio-economic and cultural backgrounds.

Respondents were asked to comment on how they believed the Auckland Tuatara and Baseball New Zealand could work more efficiently together to enhance peri- and post-event well-being outcomes. It was acknowledged that the NSB needed to have a stronger presence at the games to strengthen the links between perceived peri-event well-being outcomes and eventual post-event well-being outcomes. For example, one respondent noted that there was an opportunity to better 'emphasise role models and inspirational nature of the team and then tell people how they can become involved in the sport'. Another respondent recommended Baseball New Zealand get involved in stadium activities 'to further enhance togetherness in the stadium'. There were also recommendations for the community atmosphere created at the games to be harnessed by the NSB and 'built upon further after the Tuatara games'. Other recommendations suggested the NSB organising school groups to come to the games so that they could experience and engage in the sport activity zone and be encouraged to participate in the sport. It was noted that this would require the NSB 'working more closely with schools in the lead up to the event'.

Discussion

This case study reinforces the notion that sport events have the ability to generate a range of well-being outcomes (Mayoh & Jones, 2015). Survey respondents acknowledged the generation of physical and psychological well-being outcomes both peri and post the Auckland Tuatara home-games. The findings also reiterate the conclusions of Lee, Cornwell and Babiak (2012), in that sport events can heighten a sport's impact on society. Such impacts are discussed later in relation to specifically noted physical and psychological outcomes and the identified need to further enhance well-being outcomes through effective event leveraging.

Physical well-being outcomes attributed to the Auckland Tuatara home-games were heavily slanted towards encouraging increased participation in baseball and the general promotion of physical activity, post-event. Promoting and increasing participation in the sport was also acknowledged by respondents as supporting physical-health outcomes, such as improving fitness, highlighting the importance of physical activity and promoting a healthy lifestyle. Such views echo existing research findings that indicate sport participation leads to improved overall health (Khan *et al.*, 2012; Downward & Dawson, 2016; Edwards & Rowe, 2019) and enhance the World Health Organization's (2003) views on sport participation health-related outcomes. Participation in sport and physical activity was further aided peri-event by the inclusion of the children's sport zone (established to the left, and in view of, the diamond where the Auckland Tuatara games were taking place and offering activities including throwing, catching, running and batting).

The zone presented a supportive setting for children to participate in a range of activities that promoted positive health outcomes (Edwards & Rowe, 2019). Those children who participated in the activities on offer were perceived as experiencing general happiness and enjoyment (Downward & Rascuite, 2011; Zhou & Kaplanidou, 2018). It could further be argued that through that experience they may then be encouraged to seek similar experiences through the sport, post-event (Grabs *et al.*, 2016).

In addition to contributing to physical well-being outcomes, survey respondents also noted that the event contributed to a range of psychological well-being outcomes. These include the likes of showcasing role models, inspiring children, generating a feeling of 'togetherness in the stadium', encouraging the pursuit of dreams and bringing a diverse range of socio-economic and cultural groups together. Such responses paralleled aspects of joy, social connection, positive relations with others and general acceptance noted by the likes of Lundqvist (2011) and Kim and James (2019). Such experiences can also be linked with aspects of happiness and fulfilment, both of which are considered well-being outcomes by Lee, Cornwell and Babiak (2012). Respondents' comments on the provision of role models also support the suggestion by Mayoh and Jones (2015), in that sport events encourage spectators to identify with the players they see and engage in the sport. With reference to the sports zone, respondents noted that this area also assisted in promoting a family focus and aspects of social cohesion, peri-event. These perceptions echo the findings of Schulenkorf (2012) in relation to the events offering a chance to bring communities together and the fostering of collective identity (Djaballah, Hautbois & Desbordes, 2015; Spaaij, Oxford & Jeanes, 2016; Kim & James, 2019).

It is also in reference to the psychological well-being outcomes of the home-game events where SWB viewpoints presented by Ramchandani, Coleman and Millar (2019) are apparent. Survey respondents were asked to reflect on their emotional experiences associated with the event, present evaluative feedback and consider the personal sense of meaning associated with their event experience. The memorable aspects of the home-games related largely to the inspirational aspects for young children attending the game, along with opportunities of goal realisation through reflection of experiences at the event and the showcasing of role models. Such outcomes are also common to the research presented by Downward and Rascuite (2011) and Grabs *et al.* (2016). Emotional support was demonstrated through the social cohesion and togetherness created within the stadium, similar to the findings of Inoue *et al.* (2020).

With particular reference to the NSB and the franchise, the findings highlight potential untapped leveraging opportunities aimed at further enhancing the well-being outcomes of home-game events. This was emphasised in regard to the identified need for the NSB to work more closely with the Auckland Tuatara to achieve shared well-being goals, particularly in terms of utilising home-game events to promote grassroots participation in the sport peri-event, and this increased awareness of participation in and enjoyment of the sport post-event. Respondents' recommendations, in this sense, reiterate the need for employing a leveraging approach so as to ensure event well-being outcomes are achieved (Schulenkorf, 2012; Edwards, 2015; Bell & Daniels, 2019; Edwards & Rowe, 2019). Taks *et al.* (2017) noted that it is only upon post-reflection of an event that there is recognition of missed opportunities. This was apparent in terms of extended well-being event outcomes not necessarily fully considered by Baseball New Zealand and the Auckland Tuatara ahead of the home-game events. The research findings also emphasise the suggestion made by Robson and France (2011), whereby amateur sport needs to 'piggyback' on professional sport to extend their performance as an organisation and the overall impact.

In identifying the various psychological and physical event well-being outcomes associated with the Auckland Tuatara home-games, and appreciating the value a leveraging approach would bring to extending such event well-being outcomes, further emphasis can be placed on the well-being legacies associated with future Auckland Tuatara home-game events. With emphasis on the post-event well-being outcomes, the case study research findings support the notion presented by Duarte and O'Shea (2012), Warner and Dixon (2016) and Thomson *et al.* (2019), in that event legacies are frequently aligned with well-being outcomes and the betterment of society. This emphasises the longer-term impacts of the events (Thomson, Schlenker & Schulenkorf, 2013) and ensures that such outcomes are sustainable (Zhou & Kaplanidou, 2018).

Conclusion

This case study has contributed to the discourse on sport event well-being outcomes in a number of different ways. Firstly, by confirming that local sport events, including sporting franchise home-games, can generate a range of physical and psychological well-being outcomes. Secondly, by acknowledging that well-being outcomes can be more effectively achieved when sport event leveraging initiatives are enacted and sustainable sport event legacies are conceived. Thirdly, by highlighting the need for NSBs to work more effectively with associated sporting franchises to better achieve their community-based well-being strategic goals. In turn, a strengthened strategic relationship between a sporting franchise and associated NSB can also assist in successfully achieving and extending pre-established sporting event outcomes. Fourthly, the perceived outcomes associated with the Auckland Tuatara 2019/2020 ABL season indicate that smaller events, such as sporting franchise home-games, have the ability to create lasting legacies just as much as mega sport events.

In terms of future research, additional studies could review the relationships between other NSBs and associated sporting franchises and their joint approaches to leveraging and legacy development. As an extension of the event attendee perceptions noted in this study, a direct analysis of NSB and sporting franchise relationships could also take place. In closing, it would be interesting to revisit the Baseball New Zealand/Auckland Tuatara relationship to determine if progress has been made on event leveraging and legacy efforts that aim to further extended well-being outcomes.

References

Auckland Tuatara looking for better roster for second season in ABL. (2019, March 24). *Sunday Star Times*. Available from www.stuff.co.nz/sport/othersports/111494875/auckland-tuatara-looking-for-better-roster-for-second-season-in-abl.

Baseball: Auckland Tuatara officially unveiled as New Zealand entry to Australian Baseball League. (2018, August 27). *New Zealand Herald*. Available from www.nzherald.co.nz/sport/news/article.cfm?c_id=4&objectid=12114221.

Baseball: Auckland Tuatara qualify for ABL playoffs after win over Brisbane Bandits. (2020, January 26). *Newshub*. Available from www.newshub.co.nz/home/sport/2020/01/baseball-auckland-tuatara-qualify-for-abl-playoffs-after-win-over-brisbane-bandits.html.

Baseball New Zealand. (2019). *Baseball New Zealand 2020–2023 strategic plan*. Auckland: Baseball New Zealand.

Bell, B., & Daniels, J. (2019). Sport development in challenging times: Leverage of sport events for legacy in disadvantaged communities. *Managing Sport and Leisure*. 23 (4–6), 369–390.

Bell, J. (2005). *Doing your research project*. New York: Open University Press.

Bester, P. (2012). The business environment of sport organisations: A review. *African Journal of Business Management*. 6 (46), 11530–11535.

Brajša-Žganec, A., Merkaš, M., & Šverko, I. (2011). Quality of life and leisure activities: How do leisure activities contribute to subjective well-being? *Social Indicators Research*. 102 (1), 81–91.

Brandl-Bredenbeck, H.P., & Kampfe, A. (2012). Surveys. In: Armour, K.M., & MacDonald, D. (eds.) *Research methods in physical education and youth sport*. New York: Routledge.

Brown, K., Keast, R., Charles, M., Ryan, N., & Cairncross, G. (2010). Public-private partnerships in the promotion of sport and well-being: A governance model. In, *Proceedings of the 14th annual conference of the international research society for public management*, 7–9 April. Bern, Switzerland.

Bryman, A., & Bell, E. (2011). *Business research methods* (3rd ed.). New York: Oxford University Press.

Chalip, L., Green, C., Taks, M., & Misener, L. (2017). Creating sport participation from sport events: Making it happen. *International Journal of Sport Policy and Politics*. 9 (2), 257–276.

Coghlan, A., & Filo, K. (2013). Using constant comparison method and qualitative data to understand participants' experiences at the nexus of tourism, sport and charity events. *Tourism Management*. 35, 122–131.

Cohen, L., Manion, L., & Morrison, K. (2007). *Research methods in education* (6th ed.). New York: Routledge.

Cornelissen, S., Bob, U., & Swart, K. (2011). Towards redefining the concept of legacy in relation to sport mega-events: Insights from the 2010 FIFA World Cup. *Development Southern Africa*. 28 (3), 307–318.

Creswell, J., & Miller, D. (2000). Determining validity in qualitative inquiry. *Theory into Practice*. 39 (3), 124–130.

Djaballah, M., Hautbois, C., & Desbordes, M. (2015). Non-mega sporting events' social impacts: A sensemaking approach of local governments' perceptions and strategies. *European Sport Management Quarterly*. 15 (1), 48–76.

Downward, P., & Dawson, P. (2016). Is it pleasure or health from leisure that we benefit from most? An analysis of well-being alternatives and implications for policy. *Social Indicators Research*. 126, 443–465.

Downward, P., & Rascuite, S. (2011). Does sport make you happy? An analysis of the well-being derived from spots participation. *International Review of Applied Economics*. 25 (3), 331–348.

Duarte Alonso, A., & O'Shea, M. (2012). "You only get back what you put in": Perceptions of professional sport organizations as community anchors. *Community Development*. 43 (5), 656–676.

Dunn, M. (2014). Understanding athlete wellbeing. *Sport Health*. 32 (2), 46–47.

Edwards, M. (2015). The role of sport in community capacity building: An examination of sport for development research and practice. *Sport Management Review*. 18, 6–19.

Edwards, M., & Rowe, K. (2019). Managing sport for health: An introduction to the special issue. *Sport Management Review*. 22, 1–4.

Edwards, A., & Skinner, J. (2009). *Qualitative research in sport management*. Oxford: Elsevier.

Fernandez, M., & Walker, B. (2019). Spirituality in the stands: Sports spectatorship, spirituality, wellbeing and personality. *Journal of Spirituality in Mental Health*. DOI: 10.1080/19349637.2019.1654427.

Gournelos, T., Hammonds, J.R., & Wilson, M.A. (2019). *Doing academic research: A practical guide to research methods and analysis*. New York: Routledge.

Grabs, J., Langen, N., Maschkowski, G., & Schäpke, N. (2016). Understanding role models for change: A multilevel analysis of success factors of grassroots initiatives for sustainable consumption. *Journal of Cleaner Production*. 134, 98–111.

Gratton, C., & Jones, I. (2010). *Research methods for sport studies* (2nd ed.) New York: Routledge.

Heere, B., & James, J. (2007). Sports teams and their communities: Examining the influence of external group identities and team identity. *Journal of Sport Management*. 21, 319–337.

Inoue, Y., Wann, D., Lock, D., Sato, M., Moore, C., & Funk, D. (2020). Enhancing older adults' sense of belonging and subjective well-being through sort game attendance, team identification, and emotional support. *Journal of Aging and Health*, 1–26.

Johnson, A. (1998). Minor League baseball: Risks and potential benefits for communities large and small. *Policy Studies Review*. 15 (1), 45–54.

Kassens-Noor, E., Wilson, M., Muller, S., Maharaj, B., & Huntoon, L. (2015). Towards a mega-event legacy framework. *Leisure Studies*. 34 (6), 665–671.

Kavetsos, G., & Szymanski, S. (2010). National well-being and international sport events. *Journal of Economic Psychology*. 31, 158–171.

Khan, K., Thompson, A., Blair, S., Sallis, J., Powell, K., Bull, F., & Bauman, A. (2012). Sport and exercise as contributors to the health of nations. *The Lancet*. 380, 59–64.

Kihl, L., Babiak, K., & Tainsky, S. (2014). Evaluating the implementation of professional sport team's corporate community involvement initiative. *Journal of Sport Management*. 28, 324–337.

Kim, J., & James, J. (2019). Sport and happiness: Understanding the relations among sport consumption activities, Long- and Short-Term subjective well-being, and psychological need fulfilment. *Journal of Sport Management*. 33, 119–132.

Klassen, A., Creswell, J., Plano Clark, V.L., Clegg Smith, K., & Meissner, H. (2012). Best practices in mixed methods for quality of life research. *Quality of Life Research*. 21, 377–380.

Klein, A. (2009). The transnational view of sport and social development: The case of Dominican baseball. *Sport in Society*. 12 (9), 1118–1131.

Lee, S., Cornwell, T.B., & Babiak, K. (2012). Developing an instrument to measure the social impact of sport: Social capital, collective identities, health literacy, well-being and human capital. *Journal of Sport Management*. 27, 24–42.

Lundqvist, C. (2011). Well-being in competitive sports – The feel-good factor? A review of conceptual considerations of well-being. *International Review of Sport and Exercise Psychology*. 4 (2), 109–127.

Marks, N., Thompson, S., Eckersley, R., Jackson, T., & Kasser, T. (2011). *Sustainable development and well-being: Relationships, challenges and policy implications*. London: The Department for Environment, Food and Rural Affairs.

Mayoh, J., & Jones, I. (2015). Making well-being an experiential possibility: The role of sport. *Qualitative Research in Sport, Exercise and Health*. 7 (2), 235–252.

Misener, L., & Mason, D. (2006). Creating community networks: Can sporting events offer meaning sources of social capital? *Managing Leisure*. 11, 39–56.

O'Brien, D., & Chalip, L. (2007). Sport events and strategic leveraging: Pushing towards the triple bottom line. In: Woodside, A., & Martin, D. (eds.) *Tourism management: Analysis, behaviour and strategy*. CABI Publishing, pp. 318–338. https://doi.org/10.1079/9781845933234.0318.

Pawlowski, T., Downward, P., & Rascuite, S. (2014). Does national pride from international sporting success contribute to well-being? An international investigation. *Sport Management Review*. 17, 121–132.

Preuss, H. (2010). The conceptualisation and measurement of mega sport event legacies. *Journal of Sport and Tourism*. 12 (3–4), 207–228.

Ramchandani, G., Coleman, R., & Millar, R. (2019). The perceived influence of sport event spectatorship on subjective wellbeing. *Journal of Global Sport Management*. DOI: 10.1080/24704067.2019.1669066.

Rhodes, K. (2020). Impactful corporate social responsibility in Major League Baseball. *The Sport Journal*. 41 (2), 1–16.

Robson, S., & France, A. (2011). Comparing sports marketing of amateur team sports to professional team sports. *New Zealand Journal of Applied Business Research*. 9 (1), 47–61.

Rowe, K., Shilbury, D., Ferkins, L., & Hinckson, E. (2013). Sport development and physical activity promotion: An integrated model to enhance collaboration and understanding. *Sport Management Review*. 16, 364–377.

Schulenkorf, N. (2012). Sustainable community development through sport and events: A conceptual framework for sport-for-development projects. *Sport Management Review*. 15, 1–12.

Schulenkorf, N., & Edwards, D. (2012). Maximizing positive social impacts: Strategies for sustaining and leveraging the benefits of intercommunity sport events in divided societies. *Journal of Sport Management*. 26, 379–390.

Silverman, D. (2004). *Qualitative research: Theory, method and practice* (2nd ed.). London: Sage.

Spaaij, R., Oxford, S., & Jeanes, R. (2016). Transforming communities through sport? Critical pedagogy and sport development. *Sport, Education and Society*. 21 (4), 570–587.

Sparvero, E., & Chalip, L. (2007). Professional teams as leverageable assets: Strategic creation of community value. *Sport Management Review*. 10, 1–30.

Spracken, K., & Lamond, I. (2016). *Critical event studies*. London: Routledge.

Stake, R. (1995). *The art of case study research*. London: Sage.

Steptoe, A., & Butler, N. (1996). Sport participation and emotional wellbeing in adolescents. *The Lancet*. 347, 1789–1792.

Taks, M., Green, B., Misener, L., & Chalip, L. (2017). Sport participation from sport events: Why it doesn't happen? *Marketing Intelligence & Planning*. 36 (2), 186–198.

Thomson, A., Cuskelly, G., Toohey, K., & Kennelly, M. (2019). Sport event legacy: A systematic quantitative review of literature. *Sport Management Review*. 22 (3), 295–321.

Thomson, A., Schlenker, K., & Schulenkorf, N. (2013). Conceptualizing sport event legacy. *Event Management*. 17, 111–122.

Wann, D., Melnick, M., Russell, G., & Pease, D. (2001). *Sports fans: The psychology and social impact of spectators*. London: Routledge.

Warner, S., & Dixon, M. (2016). Connecting it all: Creating community in sport and entertainment. *Sport and Entertainment Review*. 2, 45–50.

World Health Organization. (2003). *Health and development through physical activity and sport*. Geneva, Switzerland: World Health Organization.

Yancey, A., Winfield, D., Larsen, J., Anderson, M., Jackson, P., Overton, J., Wilson, S., Rossum, A., & Kumanyika, S. (2009). "Live, learn and play": Building strategic alliances between professional sports and public health. *Preventative Medicine*. 49, 322–325.

Zhou, R., & Kaplanidou, K. (2018). Building social capital from sport event participation: An exploration of the social impacts of participatory sport events on the community. *Sport Management Review*. 21, 491–503.

4 How a sports event can impact participants' subjective well-being

The case of the Pacific Games – Samoa 2019

Stephen Pratt, Nacanieli Rika and Shavneet Sharma

Introduction

Within the research on large-scale festivals and events, sporting competitions have received a significant amount of attention in the literature (Kim *et al.*, 2015). Research into the impacts on the resident community of hosting the Olympic Games has been the focus of much research (Konstantaki & Wickens, 2010; Chen & Tian, 2015). However, other large-scale sporting events have also been evaluated including the FIFA World Cup (Swart *et al.*, 2018) and the ICC Cricket World Cup (Lorde, Greenidge & Devonish, 2011). Given the significant finances required to host events of these sizes, much of the initial research involves estimating the economic impacts of hosting these events (Fourie & Santana-Gallego, 2011; Baade & Matheson, 2016). More recently, there has been a shift in direction to examine the socio-cultural impacts of hosting large-scale events on the resident population (Fredline, Jago & Deery, 2003; Yolal *et al.*, 2016) as well as impacts of events on the physical environment (Dolles & Söderman, 2015).

This research takes a different approach to examine how a large-scale sporting event, the Pacific Games, can influence the host community's subjective well-being. The Pacific Games bring together up to 24 countries and territories to compete in a multi-sports event. The 2019 edition was held in Samoa. Using a quantitative survey instrument among Pacific Games attendees, this research assesses the extent to which perceived economic, socio-cultural, and environmental impacts as a result of this sporting event influence the attending residents' subjective well-being. This research also seeks to determine how the underlying ethical values of this sporting event impact residents' quality of life. Unlike previous research, our proposed model includes assessing the likelihood of Games attendees' intention to act in a pro-environmental way and their intention to implement a healthier lifestyle as a result of attending the Pacific Games. This research can serve as a benchmark to assess how future sporting events can motivate attendees to live healthier and more environmentally sustainable lifestyles.

Literature review

Quality of life and subjective well-being of residents

There has been increased interest by researchers in tourism and hospitality in the study of quality of life (QOL) (Ramkissoon, Mavondo & Uysal, 2018; Sirgy, 2019). While objective measures such as the unemployment rate, educational attainment and GDP give some indication of the quality of life of members of society, the concept of well-being cannot be measured solely with broad objective measures (Cummins, 2000). As such, this study will employ subjective measures to measure the well-being of individuals, as a result of hosting an international sporting event. Subjective well-being aims to assess individuals' perceptions of their lives as a whole, as well as the interactions between different areas of life (McCabe & Johnson, 2013). Subjective well-being is one of the key determinants of residents' QOL (Kim, Uysal & Sirgy, 2013). When evaluating well-being, individuals consider not only their personal well-being but also their social life, health and safety, financial life, and the well-being of the community (Vargas-Sánchez, Plaza-Meji & Porras-Bueno, 2009). This plays a crucial role in shaping residents' attitudes towards supporting tourism development (Vargas-Sanchez, Porras-Bueno & de los Ángeles Plaza-Mejía, 2011).

The concept of subjective well-being is defined as an individual's 'cognitive and affective evaluations of his or her life' (Diener & Suh, 1997, p. 191). Despite the concept of subjective well-being falling under the broader field of QOL, the two concepts are often used interchangeably (Ramkissoon, Mavondo & Uysal, 2018). Subjective well-being can be a multidimensional or unidirectional construct. Applied to events, subjective well-being can be defined as how an individual reacts emotionally to an event and their cognitive evaluation of fulfilment and satisfaction the event can bring to their life (Yolal *et al.*, 2016).

According to psychologists, for an individual to experience well-being, the need for independence, competence, and relatedness must be satisfied (Ryan & Frederick, 1997). Morgan, Pritchard and Sedgley (2015) and Naidoo and Sharpley (2016) argue that participation in tourism and tourism development contributes positively to an individual's well-being. Researchers have integrated the concept of subjective well-being in sports events (Kaplanidou, 2012; Al-Emadi *et al.*, 2017). Such events provide individuals with unique opportunities to satisfy their needs (Yolal *et al.*, 2016).

Community development

Hosting large-scale events brings about socio-economic benefits to the host country (Waitt, 2003; Gursoy & Kendall, 2006). Large sporting events stimulate economic activity, of which tourism is a part, through increased trade for local businesses and increased investment in leisure facilities and transport infrastructure (Duignan, Down & O'Brien, 2020). This results in more commercial activities within the communities (Gursoy *et al.*, 2017). Hosting sporting events can improve the destination image, drawing more tourists in the future (Heere *et al.*, 2019). There are also socio-cultural impacts of such events on the local communities (Grappi & Montanari, 2011). This includes enhancing the cohesion and pride in the community, strengthening traditional and cultural values, giving opportunities to interact with individuals from different cultures, and is a source of enjoyment and relaxation for the locals (Gursoy *et al.*, 2011). Large events also give residents opportunities to share their community development and culture with visitors (Liang, Illum & Cole, 2008). Therefore, the following hypothesis is proposed:

> **H_1**. There is a positive relationship between the perceived community development from the Pacific Games and the Pacific Games participants' subjective well-being

Economic benefits

The rapid growth of sports events has been driven predominantly for their perceived economic benefits (Bakhsh *et al.*, 2018). Hosting such events has the potential to bring about the development of the host country's infrastructure and transportation, contributing to economic growth (Gursoy & Kendall, 2006). However, residents have also been found to oppose sports events (Burbank, Heying & Andranovich, 2000). According to some residents, the economic costs of organising such events outweigh the economic benefits (Leenders, Go & Bhansing, 2015). Residents feel that the construction of facilities required to host such events is a waste of taxpayers' money (Bob & Swart, 2009). Nunkoo and Gursoy (2016) also state that some residents believe that large events can lead to the tax burden, mismanagement of public funds, and a shift of public funds to private interests. Despite these findings, most of the studies (e.g. Andereck & Vogt, 2000; Andereck *et al.*, 2005) report that residents generally have a positive perception of the economic impacts. Hence, this study proposes the following hypothesis:

> **H_2**. There is a positive relationship between perceived economic benefits of the Pacific Games and the Pacific Games participants' subjective well-being

Ethics and values of sporting events

Sporting events are often created not only to test speed, strength, and endurance but also to foster integrity, goodwill, sportsmanship, fairness, and equality. The Olympic Movement was built on these ideals (Šukys & Majauskien, 2014). These values have been encoded in the Olympic Charter (Milton-Smith, 2002). Through sport, these values can resonate among participants and spectators and spill over into other life domains, combining in a balanced whole (Parry, 2006). Host countries often hold large-scale events to build cultural and social capital while building interactions at a community level that links to universal and national values (Tomlinson & Young, 2006). Sports have the potential to reach diverse populations including groups that are marginalised (Burnett, 2019). Sporting events can provide social cohesion among different groups and can foster a common understanding and a sense of belongingness (Burnett, 2019). Sporting events allow the building of relationships across divisive barriers such as ethnicity and allow for the development of collective identity (Walseth, 2006). Sports can create closer diplomatic ties and stronger international trade connections among countries (Jarvie, 2013). According to Skinner, Zakus and Cowell (2008), sporting activities allow marginalised individuals to be part of a team and contributed to the creation of community life. Similarly, Jenny and Rhodes (2017) find sports activities lead to the creation of trust and solidarity building. Developing these character traits through sport can contribute positively to an individual's well-being. For example, Cultural diversity and resilience and Community vitality are two of the life domains used in Bhutan's Gross Happiness Index (Ura *et al.*, 2012).

H_3. There is a positive relationship between perceived values of sporting events and the Pacific Games participants' subjective well-being

Health benefits

Participation in sporting activities is known to promote health outcomes (Kim *et al.*, 2019). Sport plays a critical role in improving health and productivity. Organisations have promoted sports participation as a means of improving physical health (Chalip, 2011). Lee, Pope and Gao (2018) found that youth sport participation is positively associated with children's physical activity. Another study by Kim *et al.* (2019) found that sports participation leads to various positive health outcomes such as decreased anxiety, depression, and enhanced mood state, and enhanced physical health.

H_4. There is a positive relationship between perceived health benefits of the Pacific Games and the Pacific Games participant's subjective well-being

Pro-environment behaviour

Pro-environment behaviour is a key issue of concern for residents of host countries who hold major sporting events (Han, Nelson & Kim, 2015). This particularly applies to residents that are more conservation-oriented. Such individuals consider the costs and benefits of development strategies on the environment (Gursoy *et al.*, 2019). People-environment congruity is essential for a high QOL (Moser, 2009). The staging of events can help preserve the environment (Backman *et al.*, 1995). However, there have been mixed empirical results found in terms of the influence of pro-environmental behaviour and subjective well-being. The Gursoy and Kendall (2006) study on the impacts of the 2002 Winter Olympics found that residents' ecocentric attitudes significantly influenced their perceived impact of the event. Similar results were found by Gursoy, Jurowski and Uysal (2002) and

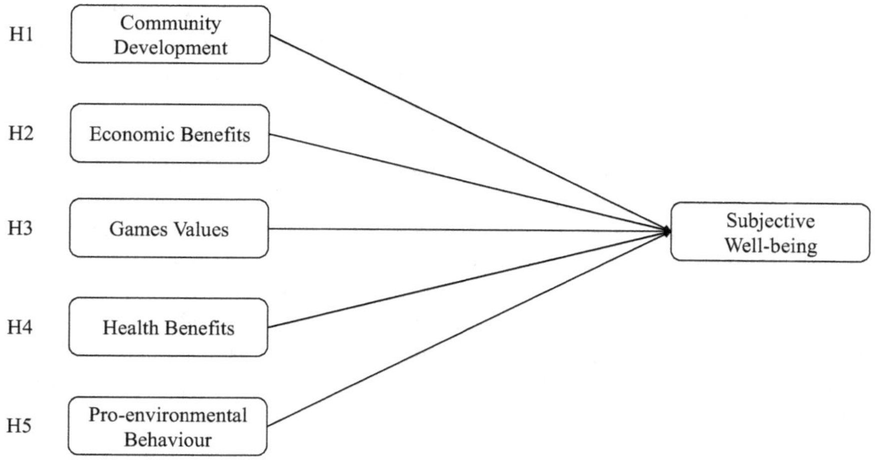

Figure 4.1 Impact of events on subjective well-being

Jurowski, Uysal and Williams (1997). Based on the earlier discussion, the following hypothesis is proposed:

> **H$_5$**. There is a positive relationship between participants' pro-environment behaviour as a result of the Pacific Games and the Pacific Games participants' subjective well-being

Based on the earlier hypotheses, the following conceptual model is developed (Figure 4.1).

Pacific Games – Samoa 2019

The Pacific Games, like the Olympics, is held exclusively for countries and territories in the Pacific Ocean. Six countries or territories, across nine cities, have hosted the Pacific Games. Four countries (Fiji: 1963, 1979, 2003; New Caledonia: 1966, 1987, 2011; Papua New Guinea: 1969, 1991, 2015; and Samoa: 1983, 2007, 2019) have been three-time hosts. In 2019, Solomon Islands was announced as the 2023 Pacific Games host becoming the seventh country to host the Games. The official languages of the Pacific Games are English and French, a legacy of many Pacific Islands Countries and Territories' colonial past. The Pacific Games Council, the Games governing body, has approved 37 sports, with 16 core sports being contested at every Games. This includes Pacific-orientated sports such as Rugby 7s and Outrigger canoeing (*va'a*).

The 2019 edition was held in Samoa over two weeks from 7 to 20 July and included 3,500 athletes from 24 nations competing in 26 sports. The 13th Pacific Games were initially awarded to Tonga who subsequently withdrew amid concerns hosting the Games would put an economic strain on this Pacific Kingdom. Samoa was subsequently chosen to be hosts in a three-way contest with Guam and French Polynesia. At the competition's end, New Caledonia had the highest medal tally with 182, of which 76 were Gold medals. The hosts, Samoa, ranked third in the medal tally, picking up 125 medals, 38 of which were Gold.

For environmentally vulnerable Pacific Island Countries like Samoa, waste management is an important issue (Hussey, 2015). Hence, the Games organisers, in conjunction with the Secretariat of the Pacific Regional Environment Programme (SPREP), implemented a 'Greening of the Games' campaign aimed at reducing single-use plastic and improving waste management. While at the Games, athletes and officials were provided with reusable bags and water bottles (Pacific Games Office, 2019). Free water stations were provided at all Games venues. Special bins were provided to separate waste at Games venues. Athletes and officials were encouraged to say 'no' to Styrofoam packaging for food and single-use plastics. Clean-ups were organised in the capital city of Apia for residents and schoolchildren.

The Pacific Games is an appropriate context to understand how hosting and participating in large-scale sporting events can impact subjective well-being. For Small Island Developing State like Samoa, this was a large event lasting two weeks with a lot of publicity and preparations leading into the hosting of the Games.

Methodology

To achieve the research objectives, the most appropriate methods to use are a quantitative study. The target population for the study is Samoan residents who participated in the Pacific Games, as athletes, officials, or spectators at events.

Survey instrument

The survey instrument comprises three sections. The first section of questions asked a standard set of demographic questions such as age, gender, marital status, highest education level, and residential status. The second section asked questions about Pacific Games participants – if they attended the Samoa 2019 Pacific Games and in what capacity, and the number and type of events they attended. The last section asked about the perceived impacts of the events and subjective well-being. The subjective well-being of residents construct was captured using three items adapted from Yolal *et al.* (2016). The community development construct included in this research adapted the five items from Kim *et al.* (2015). Additionally, the four items used by Kim *et al.* (2015) to capture economic benefits were also used in this research. Items about community development and economic benefits were asked on a 7-point Agreement Likert scale where '7' was 'Strongly Agree' and '1' was 'Strongly Disagree', as per Kim *et al.* (2015). To assess the Pacific Games' values, the researchers adapted from the Pacific Games Council Charter, specifically Article 2.6 (Pacific Games Council, 2006). Five items were asked of attendees concerning the extent that the Pacific Games achieved the following objectives: creates bonds of friendship between people of different Pacific Countries; promotes fair play; promotes respect for gender equality and tolerance; promotes a commitment to excellence; and encourages sporting exchanges without discrimination. Games Values were asked on a 5-point Agreement Likert scale where '5' was 'Strongly Agree' and '1' was 'Strongly Disagree'.

Four items were asked about intended health benefits as a result of the Pacific Games. These items cover diet (start eating more healthily) and physical activity (join a sports team or club, join a gym, do more exercise). Residents were also asked the extent to which the Pacific Games were likely to encourage pro-environmental behaviour. In line with the 'Greening of the Games' message, respondents were asked how likely they were to (1) stop using single-use plastics and polystyrene; (2) plant a tree to offset your carbon footprint; (3) use reusable water bottles and shopping bags; and (4) dispose of any waste in the provided waste and recycling bins. Health benefits and pro-environmental behaviour were asked on a 5-point Likelihood Likert scale where '5' is extremely likely and '1' is extremely unlikely. The questionnaire was designed in English and translated into Samoan. Respondents were given the option of which language they preferred to answer the survey.

Data collection

The questionnaire was programmed into SurveyMonkey, an online survey platform, and data were collected electronically via face-to-face interviews captured on tablets. The residents' data were collected at Games venues and commonly frequented locations around Samoa, such as the Central Bus Stand. This ensured that the data captured covered those attending the Games. The data collection commenced at the end of the first week of the Pacific Games and finished 2 days after the Closing Ceremony. This allowed Samoan residents a chance to have experienced the Games. A team of 12 research assistants was trained in best-practice interviewing techniques. Research assistants approached potential respondents, introduced themselves, and informed the potential respondent about the purpose of the research, informed them that responses would be aggregated and no one individual response could be identified. Potential respondents were informed they could discontinue the interview at any time. At the end of the data collection period, the sample size for Samoan residents attending the Games was $N = 334$, resulting in a maximum sampling error of ±5.3%.

Table 4.1 presents the sample profile. There is an even gender split. The sample skews relatively young with over a third of respondents (38.7%) falling into the 18–24 years category. This is to be expected as sports appeal to a younger audience. Almost four in five respondents (80.8%) are tertiary educated. Over half of the sample (57.4%) is employed, either in the formal or informal sector, and another quarter (24.0%) are either full-time or part-time students.

Those in the sample include Pacific Games spectators (54.1%), volunteers (30.3%), and athletes and officials. A wide range of sports was attended out of the 29 sports on offer. On average, attendees witnessed 3.34 different sports. Over a third (35.9%) only attended one sport but 25.1% attended four or more sports. The

Table 4.1 Sample profile

Sex		*Events Attended*	
Male	50.5%	Athletics	34.1%
Female	49.5%	Rugby 7s	27.6%
Age		Volleyball	26.0%
18–24 years	38.7%	Weightlifting	25.4%
25–29 years	26.9%	Basketball 5×5	22.6%
30–34 years	14.5%	Beach Volleyball	18.9%
35–44 years	13.6%	Archery	16.7%
45 years or older	6.3%	Tennis	15.5%
Marital Status		Swimming	15.2%
Single/Never married	59.5%	Netball	14.9%
Married/De facto	37.8%	Table Tennis	14.9%
Widowed/Divorced/Separated	2.7%	Rugby League 9s	12.7%
Current Activity		Basketball 3×3	11.1%
Employee (formal/informal)	57.4%	Football	11.1%
Student (full-time/part-time)	24.0%	Cricket	9.9%
Self-employed	8.1%	Lawn Bowls	8.0%
Employer (formal/informal)	4.8%	Badminton	7.1%
Looking for work/Unemployed	3.9%	Squash	6.8%
Subsistence for sale	2.7%	Touch	6.5%
Domestic duties	1.8%	Va'a	5.3%
Subsistence for consumption	1.5%	Powerlifting	4.6%
Retired	0.6%	Taekwondo	4.6%
Games Attendee		Open Water Swimming	4.3%
Pacific Games Spectator	54.1%	Judo	3.7%
Pacific Games Volunteer	30.3%	Boxing	3.4%
Pacific Games Athlete or Official	10.8%	Shooting	2.5%
Other	4.8%	Golf	2.2%
		Sailing	2.2%
		Triathlon	1.2%

three most commonly attended events were athletics, rugby 7s, and volleyball. Respondents were frequent attendees at the Games, attending an average of 5.5 days out of the 14 days the Games were held.

Data analysis

To test the hypotheses outlined in the Literature Review section, we use Partial Least Squares Structural Equation Modelling (PLS-SEM) using the SmartPLS software (Ringle, Wende & Becker, 2015). While covariance-based Structural Equation Modelling has been more commonly used in the past, the variance-based PLS-SEM has become increasingly preferred to test causal relationships.

Results

Table 4.2 presents the descriptive statistics of the indicators used in the model. Cronbach's alpha, a measure of internal reliability, for each of the constructs is

Table 4.2 Descriptive statistics

	Cronbach's Alpha	*Mean*	*Standard Deviation*	*Skewness*	*Kurtosis*
Subjective Well-being	**0.849**				
I'm really glad I went to the Pacific Games		6.61	0.819	−3.29	14.97
The Pacific Games has enriched my life		6.32	0.984	−2.23	7.25
The Pacific Games was rewarding to me in many ways		6.27	1.049	−1.98	5.25
Community Development	**0.874**				
Improved the image of Samoa		6.58	0.698	−2.10	6.66
Increased opportunity to inform Samoans about the rest of the Pacific		6.38	0.949	−2.44	8.13
Increased interest in international sport events		6.35	0.923	−1.93	4.69
Increased the understanding of the other cultures and societies of visitors		6.31	0.968	−2.12	6.16
Enhanced media visibility about sports		6.29	1.047	−2.13	5.54
Economic Benefits	**0.918**				
Increased trade for local businesses		6.50	0.953	−2.70	8.69
Improved economic conditions		6.45	0.987	−2.28	5.56
Increased leisure facilities		6.45	0.929	−2.22	5.60
Increased community development investments		6.45	0.935	−2.09	4.55
Games Values	**0.854**				
The Pacific Games creates bonds of friendship between people of different Pacific Countries		4.84	0.635	−5.04	26.62
The Pacific Games promotes a commitment to excellence		4.80	0.666	−4.22	19.16
The Pacific Games encourages sporting exchanges without discrimination		4.71	0.853	−3.34	10.77
The Pacific Games promotes fair play		4.61	0.999	−2.79	6.85
The Pacific Games promotes respect for gender equality and tolerance		4.53	1.103	−2.46	4.83
Health Benefits	**0.808**				
Do more exercise		4.66	0.900	−2.94	8.11
Start eating more healthily		4.60	0.943	−2.63	6.36
Join a sports team or club		4.41	1.171	−2.04	2.97
Join a gym		4.40	1.222	−2.01	2.68

(*Continued*)

Table 4.2 (Continued)

	Cronbach's Alpha	*Mean*	*Standard Deviation*	*Skewness*	*Kurtosis*
Pro-environmental Behaviour	**0.774**				
Dispose of any waste in the provided waste and recycling bins		4.78	0.685	−3.84	15.74
Use reusable water bottles and shopping bags		4.68	0.821	−3.00	9.11
Stop using single-use plastics and polystyrene		4.58	1.054	−2.60	5.60
Plant a tree to offset your carbon footprint		4.54	0.989	−2.41	5.20

all above the 0.7 level usually used as the criteria. The lowest Cronbach's alpha is 0.774 for the pro-environmental behaviour construct, well above the benchmark. The mean scores of the indicators are all very high. On a scale of 1 to 7, the indicators of subjective well-being range from 6.27 to 6.61, while on the same rating scale, the mean scores for the items for Community Development range from 6.29 for 'Enhanced media visibility about sports' to 6.58 out of 7 for 'Improved the image of Samoa'. Similarly, the items for perceptions of Economic Benefits provided by the Pacific Games also have high mean scores ranging from 6.45 to 6.50 across the four indicators.

The indicators for Games Values, Health Benefits, and Pro-environmental behaviour are all measured on a 5-point Likert scale. These items also have high mean scores. The mean scores for Games Values range from 4.53 for 'The Pacific Games promotes respect for gender equality and tolerance' to 4.84 for 'The Pacific Games creates bonds of friendship between people of different Pacific Countries'. The Pacific Games encourages participants to 'Do more exercise' (4.66), 'Start eating more healthily' (4.60), 'Join a sports team or club' (4.41), and to a lesser extent 'Join a gym' (4.40). The Pacific Games also encourages participants to undertake pro-environmental behaviour from 'Dispose of any waste in the provided waste and recycling bins (4.78) to 'Plant a tree to offset your carbon footprint' (4.54).

As can be seen from the Skewness and Kurtosis values for all the indicators, data are very positively skewed, justifying the use of PLS-SEM.

Assessment of the measurement model

We first take an evaluation of the measurement model by examining the reliability of the construct scores, indicator reliability, convergent validity, discriminant validity, and multicollinearity. Tables 4.3 and 4.4 present this information. Reliability of the construct scores is shown by the Dijkstra–Henseler's ρ_A which reveals whether the construct scores reliably represent the underlying construct.

Table 4.3 Measurement model evaluation

	ρ_A	*Composite Reliability*	*AVE*	*VIF*	*Loadings*		*Weights*	
Subjective Well-being	**0.859**	**0.851**	**0.657**					
The Pacific Games has enriched my life				2.218	0.899	***	0.423	***
The Pacific Games was rewarding to me in many ways				2.539	0.797	***	0.375	***
I'm really glad I went to the Pacific Games				1.800	0.726	***	0.342	***
Community Development	**0.883**	**0.868**	**0.574**					
Increased the understanding of the other cultures and societies of visitors				2.050	0.908	***	0.297	***
Increased interest in international sport events				2.240	0.742	***	0.243	***
Increased opportunity to inform Samoans about the rest of the Pacific				2.104	0.764	***	0.250	***
Enhanced media visibility about sports				1.803	0.777	***	0.254	***
Improved the image of Samoa				2.141	0.556	***	0.182	***
Economic Benefits	**0.923**	**0.918**	**0.737**					
Increased trade for local businesses				2.525	0.771	***	0.251	***
Improved economic conditions				3.260	0.875	***	0.285	***
Increased leisure facilities				3.025	0.948	***	0.309	***
Increased community development investments				3.328	0.831	***	0.271	***
Games Values	**0.870**	**0.847**	**0.535**					
The Pacific Games creates bonds of friendship between people of different Pacific Countries				1.881	0.484	***	0.169	***
The Pacific Games promotes fair play				1.621	0.777	***	0.271	***
The Pacific Games promotes respect for gender equality and tolerance				1.738	0.870	***	0.303	***
The Pacific Games promotes a commitment to excellence				2.544	0.649	***	0.226	***
The Pacific Games encourages sporting exchanges without discrimination				2.416	0.810	***	0.283	***
Health Benefits	**0.809**	**0.805**	**0.509**					
Stop using single-use plastics and polystyrene				1.459	0.659	***	0.318	***

(*Continued*)

Table 4.3 (Continued)

	ρ_A	*Composite Reliability*	*AVE*	*VIF*	*Loadings*		*Weights*	
Plant a tree to offset your carbon footprint				1.548	0.891	***	0.429	***
Use reusable water bottles and shopping bags				1.496	0.608	***	0.293	***
Dispose of any waste in the provided waste and recycling bins				1.528	0.514	***	0.248	***
Pro-environmental Behaviour	**0.806**	**0.770**	**0.466**					
Start eating more healthily				1.502	0.751	***	0.332	***
Join a sports team or club				1.764	0.661	***	0.292	***
Join a gym				1.768	0.772	***	0.341	***
Do more exercise				1.898	0.662	***	0.293	***

***$p < 0.01$; AVE = Average Variance Explained

Table 4.4 Heterotrait–Monotrait ratio (HTMT)

	Community Development	*Economic Benefits*	*Games Values*	*Health Benefits*	*Pro-environmental Behaviour*	*Subjective Well-being*
Community Development						
Economic Benefits	0.554					
Games Values	0.311	0.294				
Health Benefits	0.516	0.395	0.280			
Pro-environmental Behaviour	0.400	0.395	0.500	0.565		
Subjective Well-being	0.466	0.421	0.342	0.683	0.493	

The criteria cut-off is $\rho_A > 0.707$. Table 4.3 reveals that all the $\rho_A > 0.707$ meaning that more than 50% of the variance in the construct scores can be explained by the underlying latent variable. Indicator reliability can be observed with the loadings given in Table 4.3. Factor loadings are generally >0.7. Some loadings are less than 0.7 but all are significant at the 99% level of confidence. Convergent validity is evaluated by the Average Variance Explained (AVE) statistics with the cut-off usually set at 0.5 meaning more than 50% of indicators' variance is explained by the underlying latent variable. Table 4.3 shows that five of the six constructs are greater than 0.5 with the Pro-environmental Behaviour construct's AVE being marginally lower than the cut-off (0.466). Discriminant validity examines the extent to which the constructs are different from one another. Table 4.4

shows the Heterotrait–Monotrait Ratio (HTMT) for the measurement model. The accepted cut-off is HTMT <0.85. Table 4.4 shows that the values are all less than 0.85 meaning the model has good discriminant validity. Multicollinearity can be examined through the Variance Inflation Factor (VIF). The cut-off is often set at 5 as a measure of problematic multicollinearity. Table 4.3 shows that all the VIF values are less than 5 and most are less than 3 demonstrating multicollinearity is not a problem. Lastly, we can evaluate the significance of the Weights. Table 4.3 reveals that all the weights are significant at the 99% level of confidence. These weights can be thought of as regression coefficients.

Assessment of the structural model

Having established the measurement model provides a good fit of the model, showing internal consistency and reliability, convergent reliability, and discriminant reliability with little multicollinearity and significant indicators, we examine the structural model. The R^2 can be interpreted as in regression analysis showing the variance explained of the dependent variables by all the independent variables. The indices show reasonable explanatory power with health benefits being the best explained by its indicators (Table 4.5).

Table 4.6 presents the path analysis. All of the variables are significant at the 99% level of confidence. Therefore, all of the hypotheses outlined earlier are supported. In order of magnitude, health benefits have the largest impact followed by

Table 4.5 Structural model evaluation

	R^2	*Adjusted* R^2	f^2
Community Development	0.215	0.213	0.274
Economic Benefits	0.170	0.167	0.205
Games Values	0.124	0.121	0.141
Health Benefits	0.474	0.473	0.902
Pro-environmental Behaviour	0.251	0.249	0.335

Table 4.6 Structural model

	Path Coefficients	*Lower*	*Upper*	*p-value*	*Conclusion*
Subjective Well-being -> Community Development	0.454	0.300	0.630	0.000	Supported
Subjective Well-being -> Economic Benefits	0.403	0.271	0.554	0.000	Supported
Subjective Well-being -> Games Values	0.352	0.227	0.541	0.000	Supported
Subjective Well-being -> Health Benefits	0.682	0.565	0.810	0.000	Supported
Subjective Well-being -> Pro-environmental Behaviour	0.500	0.348	0.646	0.000	Supported

pro-environmental behaviour, community development, and economic benefits, and Games Values. In conclusion, there is strong evidence that the Pacific Games has contributed to the subjective well-being across a wide range of life domains, from the economic to the socio-cultural and the environmental. Importantly, this sports event has encouraged participants to live healthier lifestyles and be more conscious of the physical environment they live in.

Discussion

In the following section, we compare and contrast the results of this research with previous literature and comment more broadly on the findings.

Subjective well-being

The results indicate that the Games enhanced participants' subjective well-being. They affirm findings from previous studies in the literature on tourism and tourism development (Morgan, Pritchard & Sedgley, 2015; Naidoo & Sharpley, 2016) as well as large sporting events including the Olympics (Kaplanidou, 2012) and the Football World Cup (Kaplanidou, 2012; Al-Emadi *et al.*, 2017). Of the three items in this construct, the extent to which the games enriched participants' lives had the highest regression weight in the PLS-SEM. However, in terms of mean scores, the highest level of agreement related to participants' happiness that they had attended the Games, followed by their feeling of being enriched and rewarded in various ways. The following quote demonstrates the impact the Games had on the participant's quality of life.

> Use these games to challenge the upcoming youth to showcase their sporting talents. In order to bring positive social solidarity towards the community, economy, and nation. In a way to increase healthy conduct of good and positive lifestyle and peace.
>
> (#54, Female, 30–34 years, Volunteer)

Health benefits

Of the five factors examined in this study, the perceived health benefits of the Pacific Games exhibited the greatest impact on participants' subjective well-being. In general, the Games appear to have encouraged participants to adopt healthier lifestyles. These results are in line with the extant literature on the positive health outcomes of sport (Chalip, 2011; Kim *et al.*, 2019). Of the four items in this category, joining a gym is the strongest predictor of health benefits. Participants' agreement was slightly higher for undertaking more exercise and eating more healthily as opposed to joining a sports club, team, or gym. Games participants are more likely to undertake healthy activities that can be undertaken more easily and can be made immediately without significant expenditure. Exercise only requires an investment of time whereas joining a sports club, team, or gym is likely to involve expenditure on membership fees, sporting equipment, and transport. On the other hand, adopting a healthier diet might actually generate financial savings if participants substitute locally grown fruits and vegetables in place of imported processed food. The fact that changes to exercise and diet can be made in relative private may be an important consideration for people who do not feel confident enough to join a sporting club, gym, or team. Exercise and dietary changes can be adopted in the short term and may lead to membership in a sporting club, gym, or team in the long term.

Pro-environmental behaviour

Participants' pro-environmental behaviour had the second-largest impact on their subjective well-being. Overall, the Games appear to have encouraged participants to be more conscious of – and responsible towards – their physical environment. These results are in line with Gursoy and Kendall (2006) and Jurowski, Uysal and Williams (1997). Of the four items in this category, the strongest predictor of pro-environmental behaviour is planting a tree to offset personal carbon emissions. However, this variable yielded the lowest average score among participants. This could be explained by the fact that tree planting is likely to involve more time and money than other pro-environmental actions. To plant a tree, a person might need to purchase it and then identify a suitable location in which to plant it. Participants' average agreement was highest for disposal of waste in the receptacles provided, followed by the use of reusable water bottles and shopping bags and then shunning single-use plastics and polystyrene. Again, actions that are relatively easy and inexpensive to do are more likely to be successful. Games organisers were able to make it easy to undertake some of these behaviours. Firstly, waste receptacles and recycling bins were provided at all Games venues. Secondly, reusable water bottles and shopping bags were freely distributed to athletes during the Games. Thirdly, single-use plastics and polystyrene containers were banned at the Games, and vendors were prohibited from using them. Several Games participants felt the environmental aspect was the best part of the Games, typified by this comment:

> Best part of the Games is the Green theme, promoting a more clean environment.
>
> (#67, Female, 25–29 years, Volunteer)

Community development

Perceived community development from the Pacific Games had the third-largest impact on participants' subjective well-being. Consistent with previous studies, (Heere *et al.*, 2019; Duignan, Down & O'Brien, 2020) the results indicate that the Games produced positive outcomes in terms of community development. The strongest predictor of perceived community development was increased understanding of visitors' cultures and societies. Participants agreed that the Games had: improved Samoa's image; informed Samoans about the rest of the Pacific;

increased understanding of other cultures and societies; increased interest in international sporting events; and enhanced media visibility about sports. Surprisingly, media visibility of sport generated the least agreement. This may reflect how the media messaging around the Games focused more explicitly on culture and national pride rather than sport. As a case in point, the Games slogan was 'One in Spirit' which may have been interpreted as focusing on friendship and unity. Similarly, the Samoan Ministry of Education, Sport, and Culture (MESC) implemented a comprehensive programme whereby every school in Samoa adopted one visiting country to the Games. This programme involved parents and the community in addition to students and teachers. The Adopt-a-Country programme was integrated into the school curriculum and explicitly encouraged students to acquire new skills in art, dance, and gardening but not in sport.

Economic benefits

Perceived economic benefits from the Pacific Games had the fourth-largest impact on participants' subjective well-being. These results are consistent with several previous studies (Andereck & Vogt, 2000; Andereck *et al.*, 2005) which highlight the perceived positive economic impact of large sporting events. The strongest predictor of perceived economic benefits was an increase in leisure facilities. Participants generally agreed that the Games had resulted in increased trade for local businesses, improved economic conditions, increased leisure facilities, and increased investment in community development. There was surprisingly little difference in the average scores across all four items in this category. The lack of variation could indicate that many of the participants were not aware of the specific economic benefits from the Games because they were not affected or involved directly. However, the open-ended feedback indicated that some participants were cognisant of specific benefits. For instance:

> The infrastructure will only increase opportunities for future events here thus increase revenue and tourism for Samoa.
>
> (#96, Female, 40–44 years, Spectator)

The results could also reflect a halo effect whereby a perceived positive impact in one area was applied to other areas as well. For instance, one participant simply stated:

> The Pacific Games have provided more benefits than cost and kept the Pacific Spirit alive.
>
> (#328, Male, 30–34 years, Spectator)

Games Values

In general, participants agreed that the Pacific Games Values were realised during their 13th edition held in Samoa in July 2019. The strongest indicator of Games values related to their promotion of respect for gender equality and tolerance. However, this indicator also recorded the lowest mean score among participants. This may be because it may be challenging for the Games to promote gender equality since most of the events involve a binary divide whereby male and female athletes compete separately. Participants recorded the strongest agreement for enhanced friendship and commitment to excellence. The Games' role in promoting friendship and unity among people of different Pacific countries was aptly reflected in the following statement by a survey participant:

> The Pacific Games is an amazing opportunity for countries of the Pacific to join together and get to know more about each other. I believe in the power of unity as it will bring peace and an opportunity that we understand each other's differences. Besides sporting events, I humbly request that we do conduct other activities [so] that our people get to interact and socialise with people from other islands and hopefully we would have another chance to host the next Pacific Games in the upcoming years. And by then we should have activities such as a festival or social get together where we all get the opportunity to really know our Pacific brothers and sisters. Other than that PG 2019 has been incredible and an amazing experience for us all.
>
> (#81, Female, 25–29 years, Volunteer)

Conclusions

The Samoa 2019 Pacific Games has contributed to participants' subjective well-being across health, environmental, and socio-economic domains. While this chapter supports the findings of previous studies from the literature on tourism and sport, it also provides new evidence from a different context and region; the Pacific Games is a regional event that operates on a significantly smaller budget and scale than the Olympic Games or FIFA World Cup.

Theoretical implications

This study extends the literature on the socio-economic impacts of sport by specifically examining perceptions around the values of the Pacific Games and their impact on the subjective well-being of participants. While previous studies have examined the socio-economic impact of large sporting events such as the Olympic Games, they have largely overlooked values and ethics. This is somewhat surprising since such events are generally premised on an explicit set of values including friendship, respect, and excellence. Values are equally important for all Pacific people, including the host nation Samoa. In general, participants agreed that the Samoa 2019 Pacific Games lived up to their values, particularly friendship between people from different countries. These values also had a statistically significant impact on participants' subjective well-being.

Practical implications

This study supports the view that major sporting events can be effective in promoting various messages or themes, including those that are not directly related to sport. These results may inform policy development relating to future Games and events. Games organisers, as well as national governments and other stakeholders, can use these events to promote key messages with positive socio-economic benefits for the entire population. In particular, the results suggest that the Samoa 2019 Pacific Games was very effective in eliciting positive attitudes around the environment and community, which were specifically targeted through explicit initiatives.

While the participants' perceptions were generally positive, their open-ended feedback indicated a few areas of concern, which may be considered by governments and the organisers of future Games. These include suggestions to promote and invest in certain sports such as *va'a*, which is a uniquely Pacific form of open-sea canoe racing, and indoor swimming, which is hampered by the lack of suitable facilities in many Pacific countries. A few participants also highlighted issues relating to the participation of transgender athletes, which was a controversial issue in the Samoa 2019 Pacific Games. This indicates the need for clearer communication and better education around transgender participation in sport.

Areas for future research

This study presents some interesting areas for future research. A follow-up survey could be conducted one year after the Samoa 2019 Pacific Games to assess their long-term impacts, particularly in terms of health benefits and pro-environmental behaviour, which rated strongly in the survey. The follow-up study could examine the extent to which Samoans have continued to exercise and eat more healthily as well as the proportion who have joined a sporting club, team, or gym as a result of the Games. Such a study would be useful in teasing out the long-term impact or legacy of major sporting events.

References

Al-Emadi, A., Kaplanidou, K., Diop, A., Sagas, M., Le, K., & Al-Ali Mustafa, S. (2017). 2022 Qatar world cup: Impact perceptions among Qatar residents. *Journal of Travel Research*. 56 (5), 678–694.

Andereck, K.L., Valentine, K.M., Knopf, R.C., & Vogt, C.A. (2005). Residents' perceptions of community tourism impacts. *Annals of Tourism Research*. 32 (4), 1056–1076.

Andereck, K.L., & Vogt, C.A. (2000). The relationship between residents' attitudes toward tourism and tourism development options. *Journal of Travel Research*. 39 (1), 27–36.

Baade, R.A., & Matheson, V.A. (2016). Going for the gold: The economics of the Olympics. *Journal of Economic Perspectives*. 30 (2), 201–218.

Backman, K.F., Backman, S.J., Uysal, M., & Sunshine, K.M. (1995). Event tourism: An examination of motivations and activities. *Festival Management and Event Tourism*. 3 (1), 15–24.

Bakhsh, J., Potwarka, L.R., Nunkoo, R., & Sunnassee, V. (2018). Residents' support for the Olympic games: Single host-city versus multiple host-city bid arrangements. *Journal of Hospitality Marketing & Management*. 27 (5), 544–560.

Bob, U., & Swart, K. (2009). *Resident perceptions of the 2010 FIFA soccer world cup stadia development in Cape Town*. Paper presented at the Urban Forum.

Burbank, M.J., Heying, C.H., & Andranovich, G. (2000). Antigrowth politics or piecemeal resistance? Citizen opposition to Olympic-related economic growth. *Urban Affairs Review*. 35 (3), 334–357.

Burnett, C. (2019). Framing a 21st century case for the social value of sport in South Africa. *Sport in Society*, 1–16.

Chalip, L. (2011). The future past of the amateur sports act: Developing American sport. *Journal of Coaching Education*. 4 (2), 4–29.

Chen, F., & Tian, L. (2015). Comparative study on residents' perceptions of follow-up impacts of the 2008 Olympics. *Tourism Management*. 51, 263–281.

Cummins, R.A. (2000). Objective and subjective quality of life: An interactive model. *Social Indicators Research*. 52 (1), 55–72.

Diener, E., & Suh, E. (1997). Measuring quality of life: Economic, social, and subjective indicators. *Social Indicators Research*. 40 (1–2), 189–216.

Dolles, H., & Söderman, S. (2015). Addressing ecology and sustainability in mega-sporting events: The 2006 football world cup in Germany. *Journal of Management & Organization*. 16 (4), 587–600.

Duignan, M.B., Down, S., & O'Brien, D. (2020). Entrepreneurial leveraging in liminoidal olympic transit zones. *Annals of Tourism Research*. 80, 102774.

Fourie, J., & Santana-Gallego, M. (2011). The impact of mega-sport events on tourist arrivals. *Tourism Management*. 32 (6), 1364–1370.

Fredline, L., Jago, L., & Deery, M. (2003). The development of a generic scale to measure the social impacts of events. *Event Management*. 8 (1), 23–37.

Grappi, S., & Montanari, F. (2011). The role of social identification and hedonism in affecting tourist re-patronizing behaviours: The case of an Italian festival. *Tourism Management*. 32 (5), 1128–1140.

Gursoy, D., Chi, C.G., Ai, J., & Chen, B.T. (2011). Temporal change in resident perceptions of a mega-event: The Beijing 2008 Olympic games. *Tourism Geographies*. 13 (2), 299–324.

Gursoy, D., Jurowski, C., & Uysal, M. (2002). Resident attitudes: A structural modeling approach. *Annals of Tourism Research*. 29 (1), 79–105.

Gursoy, D., & Kendall, K. (2006). Hosting mega events: Modeling locals' support. *Annals of Tourism Research*. 33 (3), 603–623.

Gursoy, D., Ouyang, Z., Nunkoo, R., & Wei, W. (2019). Residents' impact perceptions of and attitudes towards tourism development: A meta-analysis. *Journal of Hospitality Marketing & Management*. 28 (3), 306–333.

Gursoy, D., Yolal, M., Ribeiro, M.A., & Panosso Netto, A. (2017). Impact of trust on local residents' mega-event perceptions and their support. *Journal of Travel Research*. 56 (3), 393–406.

Han, J.H., Nelson, C.M., & Kim, C. (2015). Pro-environmental behavior in sport event tourism: Roles of event attendees and destinations. *Tourism Geographies*. 17 (5), 719–737.

Heere, B., Wear, H., Jones, A., Breitbarth, T., Xing, X., Paramio Salcines, J.L., . . . Derom, I. (2019). Inducing destination images among international audiences: The differing effects of promoting sport events on the destination image of a city around the world. *Journal of Sport Management*. 33 (6), 506–517.

Hussey, K. (2015). Climate mitigation and waste management in the Asia-Pacific. In: Calster, G.V., Vandenberghe, W., & Reins, L. (eds.) *Research handbook on climate change mitigation law*. Cheltenham, UK: Edward Elgar Publishing, pp. 407–422.

Jarvie, G. (2013). War, peace and a new world paved with good intentions through sport. Available from www.e-ir.info/2013/01/29/war-peace-and-a-new-world-paved-with-good-intentions-through-sport/ [Accessed 21 January 2020].

Jenny, S.E., & Rhodes, S. (2017). Physical education professionals developing life skills in children affected by poverty. *Physical Educator*. 74 (4), 653–671.

Jurowski, C., Uysal, M., & Williams, D.R. (1997). A theoretical analysis of host community resident reactions to tourism. *Journal of Travel Research*. 36 (2), 3–11.

Kaplanidou, K. (2012). The importance of legacy outcomes for Olympic games four summer host cities residents' quality of life: 1996–2008. *European Sport Management Quarterly*. 12 (4), 397–433.

Kim, A.C.H., Park, S.H., Kim, S., & Fontes-Comber, A. (2019). Psychological and social outcomes of sport participation for older adults: A systematic review. *Ageing & Society*, 1–21.

Kim, K., Uysal, M., & Sirgy, M.J. (2013). How does tourism in a community impact the quality of life of community residents? *Tourism Management*. 36, 527–540.

Kim, W., Jun, H.M., Walker, M., & Drane, D. (2015). Evaluating the perceived social impacts of hosting large-scale sport tourism events: Scale development and validation. *Tourism Management*. 48, 21–32.

Konstantaki, M., & Wickens, E. (2010). Residents' perceptions of environmental and security issues at the 2012 London Olympic games. *Journal of Sport & Tourism*. 15 (4), 337–357.

Lee, J.E., Pope, Z., & Gao, Z. (2018). The role of youth sports in promoting children's physical activity and preventing pediatric obesity: A systematic review. *Behavioral Medicine*. 44 (1), 62–76.

Leenders, M.A., Go, F.M., & Bhansing, P.V. (2015). The importance of the location in hosting a festival: A mapping approach. *Journal of Hospitality Marketing & Management*. 24 (7), 754–769.

Liang, Y., Illum, S.F., & Cole, S.T. (2008). Benefits received and behavioural intentions of festival visitors in relation to distance travelled and their origins. *International Journal of Event Management Research*. 4 (1), 12–23.

Lorde, T., Greenidge, D., & Devonish, D. (2011). Local residents' perceptions of the impacts of the ICC Cricket World Cup 2007 on Barbados: Comparisons of pre- and post-games. *Tourism Management*. 32 (2), 349–356.

McCabe, S., & Johnson, S. (2013). The happiness factor in tourism: Subjective well-being and social tourism. *Annals of Tourism Research*. 41 (1), 42–65.

Milton-Smith, J. (2002). Ethics, the Olympics and the search for global values. *Journal of Business Ethics*. 35 (2), 131–142.

Morgan, N., Pritchard, A., & Sedgley, D. (2015). Social tourism and well-being in later life. *Annals of Tourism Research*. 52, 1–15.

Moser, G. (2009). Quality of life and sustainability: Toward person–environment congruity. *Journal of Environmental Psychology*. 29 (3), 351–357.

Naidoo, P., & Sharpley, R. (2016). Local perceptions of the relative contributions of enclave tourism and agritourism to community well-being: The case of Mauritius. *Journal of Destination Marketing & Management*. 5 (1), 16–25.

Nunkoo, R., & Gursoy, D. (2016). Rethinking the role of power and trust in tourism planning. *Journal of Hospitality Marketing & Management*. 25 (4), 512–522.

Pacific Games Council. (2006). Pacific games council charter. Available from http://websites.sportstg.com/get_file.cgi?id=36687313 [Accessed 22 January 2020].

Pacific Games Office. (2019). Greening of the games. Available from www.samoa2019.ws/news/greening-of-the-games-have-your-say-now/ [Accessed 3 October 2019].

Parry, J. (2006). Sport and Olympism: Universals and multiculturalism. *Journal of the Philosophy of Sport*. 33 (2), 188–204.

Ramkissoon, H., Mavondo, F., & Uysal, M. (2018). Social involvement and park citizenship as moderators for quality-of-life in a national park. *Journal of Sustainable Tourism*. 26 (3), 341–361.

Ringle, C.M., Wende, S., & Becker, J.-M. (2015). SmartPLS 3. Available from www.smartpls.com [Accessed 7 December 2019].

Ryan, R.M., & Frederick, C. (1997). On energy, personality, and health: Subjective vitality as a dynamic reflection of well-being. *Journal of Personality*. 65 (3), 529–565.

Sirgy, M.J. (2019). Promoting quality-of-life and well-being research in hospitality and tourism. *Journal of Travel & Tourism Marketing*. 36 (1), 1–13.

Skinner, J., Zakus, D.H., & Cowell, J. (2008). Development through sport: Building social capital in disadvantaged communities. *Sport Management Review*. 11 (3), 253–275.

Šukys, S., & Majauskien, D. (2014). Effects of an integrated Olympic education program on adolescent athletes' values and sport behavior. *Social Behavior and Personality: An International Journal*. 42 (5), 811–821.

Swart, K., George, R., Cassar, J., & Sneyd, C. (2018). The 2014 FIFA world cup™: Tourists' satisfaction levels and likelihood of repeat visitation to Rio de Janeiro. *Journal of Destination Marketing & Management*. 8, 102–113.

Tomlinson, A., & Young, C. (2006). *National identity and global sports events: Culture, politics, and spectacle in the Olympics and the football World Cup*. Albany: SUNY Press.

Ura, K., Alkire, S., Zangmo, T., & Wangdi, K. (2012). A short guide to gross national happiness index. Available from https://opendocs.ids.ac.uk/opendocs/bitstream/handle/20.500.12413/11807/Short-GNH-Index.pdf [Accessed 11 March 2020].

Vargas-Sánchez, A., Plaza-Mejia, M.D.L.Á., & Porras-Bueno, N. (2009). Understanding residents' attitudes toward the development of industrial tourism in a former mining community. *Journal of Travel Research*. 47 (3), 373–387.

Vargas-Sanchez, A., Porras-Bueno, N., & de los Ángeles Plaza-Mejía, M. (2011). Explaining residents' attitudes to tourism: Is a universal model possible? *Annals of Tourism Research*. 38 (2), 460–480.

Waitt, G. (2003). Social impacts of the Sydney Olympics. *Annals of Tourism Research*. 30 (1), 194–215.

Walseth, K. (2006). Sport and belonging. *International Review for the Sociology of Sport*. 41 (3–4), 447–464.

Yolal, M., Gursoy, D., Uysal, M., Kim, H.L., & Karacaoğlu, S. (2016). Impacts of festivals and events on residents' well-being. *Annals of Tourism Research*. 61, 1–18.

5 A paradigmatic shift in a hallmark event extending well-being to a non-host city

"One people one nation, one Singapore"

Sandra Goh, Jagdeep Kaur Sabharwal and K. Thirumaran

Introduction

The year 2016, the joy, the cheers from the television audiences and live audiences, inexorably jubilant. Even public schools and workplaces had a heartbeat stop and appreciate the pride of the moment that would create an impact on both the emotions and later the actions of the country's citizens. Joseph Schooling, from Singapore, came in first in record-breaking time of 50.39 seconds in the 100 metres butterfly stroke. A hallmark event was born, for Singapore.

A small island nation located in Southeast Asia, Singapore has a population of little over five million citizens (Department of Statistics Singapore, 2020). Leisure sports is one of the most common sights in public parks and sports venues. There are 26 public pools in addition to private residential and fitness centre pools. Citizens entering national service are also given basic training in swimming. Singapore continues to exhibit sporting successes in regional games such as the Southeast Asian Games and Asian Games, primarily in swimming, sharpshooting, table tennis and weightlifting. In the Olympics, the country has won one silver medal in 2008 and two bronze medals in 2012 for table tennis. So, in 2016, when a swimmer from the island nation won the first gold medal in swimming, there was an air of euphoria and a sense of having arrived into the international field of competitive sports. The signification of the event leads us to use Singapore and the Olympics swimming gold medal win hosted by Brazil, as a case study to illustrate the extension of hallmark events' impact that radiates to audiences beyond the geography of the host city.

Hallmark events have a strong correlation with nationalism that often can trigger a set of events that may have a lasting impact on society. In Hall's definition, a hallmark event is located in a host country and the focus is on the event itself (Hall, 1989). Hallmark event is also related to iconic events in the sense of it being a euphoric moment to remember or a turning point for a country or people (Viol *et al.*, 2018). Stretching the definition of hallmark and iconic event as the same, this chapter explores the impact of Singapore's national swimmer, Joseph Schooling's gold medal win in 2016, 100 metres butterfly competition held in Brazil. Beyond the concept of tourists' experience at the Olympics where this event occurred, this chapter expounds on the multiplier effect on the country most impacted, Singapore. The chapter shows that such a hallmark event can reverberate in a country where reactions both online and changes in citizenry behaviour are affected as a result. As a record of analysis, the chapter uses netnographic study approach using various online sources to piece together four key premises:

1 Hallmark events can have a long-term psychological impact on the national conscious of non-host community.
2 A turning point can be triggered by a hallmark event, which can contribute to national well-being and a healthy lifestyle.
3 A copycat following the role model, which can be a motivation that can trigger a transformative response in their own life.
4 Icons in hallmark events can also transform perceptions and practices of body physique.

According to Ericsson (2002), accomplishment is defined by achievement, success or mastery at the highest level possible within a particular domain. For example, in sports, accomplishment is measured through the number of medals earned at competitions such as the Olympics. However, beyond one's personal accomplishment at the Olympics, this chapter looks at the spill over effects of the achievement on a nation's well-being and explores the aftermath of an event that resulted in a heightened interest of a nation towards sport and specifically swimming.

The study reveals that the "feel-good factor" of the host may be large and significant, although the impact of national athletic success on happiness is less significant (Kavetsos & Szymanski, 2010). A feel-good factor is derived from the satisfaction of attending events, being involved as a volunteer, being in proximity to the event without physically being there, of cultural display and national pride when a national athletic is showcased and who is also successful.

Information on the social media, which fosters communication and personal expression "constitute goldmines of information on citizens' subjective feelings" (Forgeard *et al.*, 2011, p. 98). This chapter seeks to explore the richness of these data by drawing insights from Facebook, Instagram, Twitter and YouTube, to examine the impact of the feel-good factor, and the resultant actions that promote the overall well-being of the nation.

Literature review

Scholars studying hallmark events have thus far given emphasis to the event and its impact on the immediate geographical setting (Todd, Leask & Ensor, 2017; Ouyang, Gursoy & Chen, 2019; Van der Steen & Richards, 2019; Chirieleison, Montrone & Scrucca, 2020). This chapter recognises the paradigmatic shift from studies that limit to hallmark events on the host community, tourists and economy, to evidencing the impact of a participating community that is distant and a non-host.

Ritchie (1984, p. 2) defined hallmark event as a

> [m]ajor one-time or recurring events of limited duration, developed primarily to enhance the awareness, appeal and profitability of a tourism destination in the short and/or long term. Such events rely for their success on uniqueness, status, or timely significance to create interest and attract attention.

This definition implies that there is no limit on the number of times the hallmark event can be carried out. The event, however, must make a marked contribution or be of import. Later, Hall (1989, p. 263) defines it as tourist events with the primary function to provide the host community with a competitive edge to promote their destination during a hallmark event:

> Hallmark tourist events [including] major fairs, expositions, cultural and sporting events of international status which are held on either a regular or a one-off basis. A primary function of the hallmark event is to provide the host community with an opportunity to secure high prominence in the tourism marketplace. However, international or regional prominence may be gained with significant social and environment costs.

The primary function of hallmark events and the accompanying impacts to the host cities and communities have been discussed at large (e.g. Hall, 1987, 1996; Waitt, 2003). It is time to also consider functions that impact on the participating destinations and their communities further afield. This includes stakeholders watching the event live on television screens, mobile phones and computer monitors. According to Getz (2005), a hallmark event is inextricably linked to the destination of the host city. This chapter offers another perspective to consider the legacy of hallmark event in transforming the well-being of participating nations, herein known as the non-host community and residents in this chapter.

It has been a well-known fact that hallmark sporting events such as the Olympics brought euphoria to the residents of the host city. The level of enthusiasm for the event was evidently more pronounced during the year of the game (Waitt, 2003, p. 212):

> The "buzz" surrounding the games was expressed particularly in feelings of patriotism, community spirit, and the desire to participate as a volunteer. Unquestionably, a significant psychological reward for many respondents was that the imagined bond that underpins national identity became a lived reality over the 16 days.

Although the "circus" may not be in the land of the non-host city, the "buzz" and its spill over effects on the social well-being of the non-host residents and communities of the participating countries remain unexplored.

The PERMA framework to understand well-being

Positive psychology is the "scientific study of optimal human functioning" (Linley *et al.*, 2006, p. 8). Its foundations lay in humanistic psychology and focus on positive emotion and ways to advance well-being and optimal functioning in healthy individuals (Seligman & Csikszentmihalyi, 2000; Ryan & Deci, 2001). Well-being is a multifaceted construct and includes both hedonic and eudemonic elements (Seligman, 2012; Ascenso, Perkins & Williamon, 2018; Lai *et al.*, 2018). While the hedonic element focuses on happy and pleasurable experiences (Kahneman, Diener & Schwarz, 1999; Ryan & Deci, 2001), the eudemonic element conceptualises well-being as the realisation of one's true potential (Ryff & Keyes, 1995) and the experience of purpose or meaning in life (Ryff, 1989). In positive psychology, positive well-being refers to a state where an individual is happy, highly engaged, and has a sense of purpose in life (Lee Duckworth, Steen & Seligman, 2005).

Seligman (2012) adopted a positive psychology perspective and proposed the PERMA framework, which includes elements of subjective and psychological well-being. He identified five key indicators of well-being: Positive emotions, Engagement, Relationships, Meaning, and Accomplishment (PERMA). Positive emotions are similar to hedonic happiness and include feelings such as hope, joy and contentment. Engagement occurs when an individual is highly absorbed and interested in an activity. Positive relationships include social support, social integration and satisfaction with one's social relationships. Meaning refers to feeling a sense of purpose from life and a connection to something greater than oneself. Accomplishment involves being driven to achieve one's goals and feeling self-efficacious.

The PERMA framework has been used as a well-being profiler for individuals and larger communities in different populations (Ascenso, Perkins & Williamon, 2018; Lai *et al.*, 2018). Researchers have also used positive psychology approach to examine the individual-level benefits of sports consumption (Doyle *et al.*, 2016). While a large body of research focused on the impact of sports on the participants' well-being found a positive and significant relationship (Edwards & Steyn, 2008; Williams, 2009; Rasciute & Downward, 2010; Downward & Rasciute, 2011), not many works have examined the influence of sports and sporting events on the well-being of the spectators (Doyle *et al.*, 2016). Sports spectating has become a common leisure activity (Trail, Anderson & Fink, 2000) and brings about a host of benefits to spectators including heightened well-being (Grove, Dorsch & Hopkins, 2012; Wang, Min & Kim, 2013). Choi, Kim and Kwon (2011), in their study of spectators at a major motorsport in Korea, found both eudaimonic and hedonic well-being to be positively related to spectator motives such as aesthetics and family.

The PERMA framework allows us to understand the association between sports consumption and well-being. Attending sports events is associated positively with happiness (Powlowski, Downward & Rasciute, 2014) and joy, which significantly contribute to one's well-being. Hallmann, Breuer and Kühnreich

(2013) reported that two out of three spectator respondents felt pride and happiness when German athletes were successful at major events. Watching the home team win a sporting event can lead to a vicarious sense of achievement and accomplishment, especially more so if it is an elite sporting event like Olympics (van Hilvoorde, Elling & Stokvis, 2010). Elite sport events affect national pride and public interest in sport, and national sporting performance and success could play an important role in this regard and could elevate a person's self-esteem (van Hilvoorde, Elling & Stokvis, 2010; Hallmann, Breuer & Kühnreich, 2013).

In positive psychology, engagement refers to an individual's absorption derived from interaction with an object (Seligman, 2012). When a person is engaged, they lose track of time, while completely focused on the task at hand (Seligman, 2012). According to Forgeard *et al.* (2011) engagement "refers to a psychological state in which individuals report being absorbed by and focused on what they are doing" (p. 84). This complete concentration and immersion in the experience have been referred to as "flow" (Csikszentmihalyi, 1996), a feeling of timelessness, when things just come together, and the aftermath of the flow experience is invigorating. "Flow" can lead to experiencing life more fully and intensely (Csikszentmihalyi, 2013). High levels of engagement have been linked to high levels of well-being (Kanste, 2011) since engaged persons often experience positive emotions (Bakker *et al.*, 2008). In sports, this absorption can manifest itself as degree to which the spectator is engrossed and absorbed in the sporting event. Of all the components of the PERMA framework, engagement has been the one which has proven to be most challenging to identify because of how people might recall their degree of engagement with an event in retrospect (Filo & Coghlan, 2016).

Since attending sporting events results in interactions and bonding among friends and family members (Trail, Anderson & Fink, 2000), it satisfies the "relationship" component of the framework. Meeting others at games and in the community enables people to strengthen existing, and create new, interpersonal connections (Doyle *et al.*, 2016). Sport fandom has been identified as an important avenue to gaining valuable social ties (Wann & James, 2019). Relationship is also established if spectators experience a deep connect with the sportsperson or the team. This in turn is influenced by what the sport organisations or media do to promote new or maintain existing relationships with fans and spectators (Mahony, Madrigal & Howard, 1999). In their study, Wann and Weaver (2009) found identification with a local team to be positively related to social well-being. Sports participation can be nurtured by "relationship marketing" which is defined as "attracting, maintaining and – in multi-service organisations – enhancing customer relationships" (Berry, 2002, p. 61). One can assume that successful marketing of sporting event creates an exchange relationship (Morgan & Hunt, 1994; Wang, Ho & Zhang, 2012) between the spectators and the sport and bring together a wide range of societal groups. Evidence suggests that the social connections derived from sport spectatorship reduce the likelihood that an individual will experience negative emotions (Wann *et al.*,1999) and increase the likelihood that they will experience positive feelings (Branscombe & Wann, 1991) and social connectedness (Wann, 2006).

In positive psychology, meaning is one of the key elements of well-being. Meaning is eudemonic aspect of well-being that refers to beliefs about being part of something bigger (Seligman, 2012). It allows people to understand or make sense of who they are and gives them a reason for their life (Seligman, 2012). Most direct path to happiness and well-being is through engaging in meaningful pursuits and the development of one's strengths. Meaning contributes to increased participation in charity sports events since individuals derive value and a sense of purpose from their participation in such noble causes (Coghlan & Filo, 2013; Doyle *et al.*, 2016). In sporting events, how the spectators derive the symbolic meaning will depend on how the event has been promoted or marketed (Funk & James, 2006). Wann, Hackathom and Sherman (2017) found that local sport team identification and general sport fandom positively predicted a sense of belonging and meaning in life. Though the extant literature is full of works exploring the relevance of meaning to well-being, there is a paucity of sport spectator literature conceptualising meaning within the context of PERMA framework (Doyle *et al.*, 2016).

Seligman (2012) highlights the importance of accomplishment in enabling an individual to flourish and improve their well-being. The sense of having attained can infuse a person with sense of autonomy and competence (Ryan & Deci, 2000), which facilitates optimal social and personal growth. This is more so, if the goals are close to the person's heart and they are personally invested in attaining them. Attainment of these meaningful life goals provides relatively direct satisfaction of the basic needs, thus enhancing the person's well-being (Ryan *et al.*, 1996). Sport spectatorship involves experiencing success indirectly through the sportsperson or the team. Witnessing team victories can trigger a sense of accomplishment in the spectators, more so if they feel a deep connection with the sporting team or the player (Wann, 2006; Coghlan & Filo, 2013). Hence, PERMA framework can be an apt model to understand how a non-host nation's well-being is influenced by a hallmark event such as "Olympic". The transformational outcome of a nation's well-being from an Olympic win is worthy of further study.

Method

Netnography is employed to analyse the sentiments and culture of sport, more specifically swimming in the non-host residence and community in Singapore during the Rio Olympic Games in 2016. It is a qualitative approach that draws on computer-mediated communications or network-based data, textual and visual, to develop an ethnographic understanding of an online social experience or cultural phenomenon (Kozinets, 2015). Content analysis is used to expedite the coding and analysis of data (Kozinets, 2002) and focuses on new aspects of social life (Nind *et al.*, 2012). The content from the social media is often immediate expression of the lived experience, sentiments and emotions of its user and has the potential to offer authentic information on the reactions by a particular community. The concepts of community and culture are at the core of the method.

Whalen (2018) suggests the use of different methods for examining community cultural groups from communication-based online platforms. In this study, personal reflexivity was used to recall the lived experience of a triumphant moment during the Rio Olympic Games. It is worthy to note that the researchers in this study are all residents of Singapore, who are familiar with the historical and cultural aspects of the nation's Olympic dream and culture. In-culture experience and awareness add credibility to our approach based on the reflections of readings and finding meanings as local observers. We diverge from Whalen's approach in that we were not involved in the interactions but were online observers, readers and shared an in-culture identity that added to personal and team reflexivity in this research exercise. The procedures during data analysis were otherwise like any other comparable types of qualitative interpretation (Kozinets, Dolbec & Earley, 2014).

Data collection and analysis

According to Whalen (2018), the unit of analysis should belong to a distinct community and not based on a channel with standard reviews. It is therefore essential to differentiate the community from the channel of communication, as the latter do not have a sustained cultural community or membership compared to the earlier. The rationale is that standard review websites fall short on member interactions that are usually present in the websites of identified cultural groups (Boellstorff *et al.*, 2012). Following the recommendation from Kozinets, Dolbec and Earley (2014), the criteria in Table 5.1 are observed when selecting the online communities and/or sites (entrée) to study/evaluate the premises stated in the introduction.

There was no limit on the number of comments extracted from the shortlisted sites in Table 5.2. The saturation rules were similar to that of other qualitative techniques. The collection and analysis of data stopped where there were no new insights on the research problems (Kozinets, 2015). The themes were extracted using the key indicators of PERMA as shown in Figure 5.1.

Table 5.1 Criteria for site selection/entrée

Criteria	*Action for This Study*
1 Degree of relevance to research question	• *Joseph Schooling's* social (Instagram, Twitter)
2 Higher frequency of postings	• *Joseph Schooling* Facebook Page
3 Richer data	• Olympic YouTube and News Videos
4 Presence of interactions between members	• Local media (The Straits Times, Today, Channel News Asia. . .)

Table 5.2 Response analysis of hallmark event on national well-being

Content Analysis/ PERMA	*During the Olympic*	*Post-Olympic*
Positive emotions	– Tears of joy – Motivation – Hope – Happiness – Fulfilment – Pride	– Motivation – Hope – Trust – Optimism
Engagement	– National solidarity – Celebration – Immersion in the event – "Flow"	– Huge crowd turnout for the victory parade
Relationships *TV spectators and their national athlete*	– Spectator, the fan, the follower and the "worshippers" – National heroic figure – Bonding over the event – Sense of camaraderie	– The fan, the follower and the "worshippers" – National heroic figure – Role Model – Commodifying the Olympic dream
Meaning	– Hope for future champions – The victory for the "little red dot" – A sense of purpose – Finding direction (young swimmers)	– Setting benchmarks – Changing perception of career in sports (esp. Singaporean parents) – Overseas sport education
Accomplishment	– Record-breaking – Making history – National pride – Positive self-esteem	– See identity – Cultivating young swimmers
Identity[1]	– Cultural identity – "Asian/ Eurasian Olympian" – Best moment in history	– Best moment in history – Olympic/Swim legacy, a legend – "Schooling effect" – Develop a new sense of professionalism in Team Singapore – Cultivating young swimmers

[1]Additional indicator identified during data analysis.

Findings

PERMA (during and post-Olympic games)

We were not able to identify and extract PERMA indicators in the period before the Olympics because there was a paucity of relevant online social activity. Although there were prior swimming events leading up to the Olympics, such as the SEA Games in 2015 and the FINA World Championships in 2016 in which

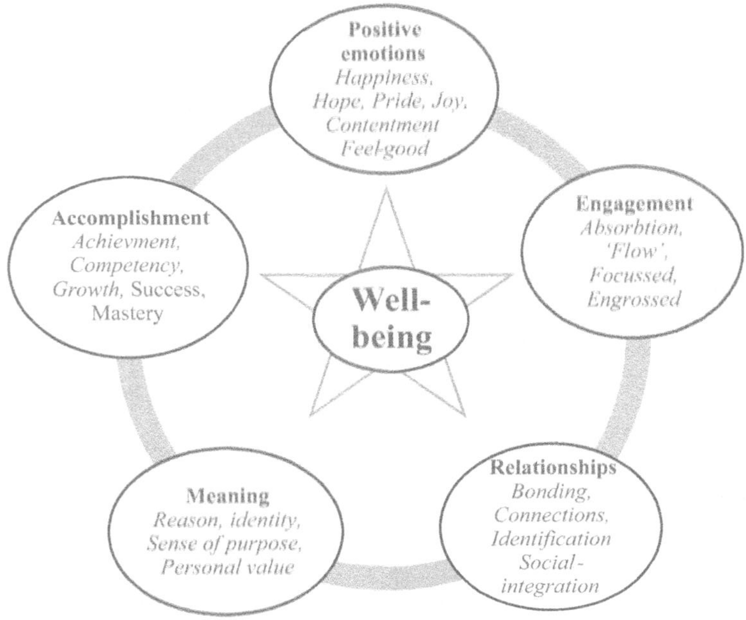

Figure 5.1 Key indicators of PERMA framework operationalised for this study

Schooling participated, these events did not bring the nation together like the 2016 Rio Olympics. For the Rio Olympics, the national swimmer, fans and supporters became one, such that the indicators on PERMA are observable during the 100 metres butterfly final on 12 August 2016. The sentiments of the spectators in Singapore were collected from fans' and supporters' responses to Joseph Schooling's win. The following findings present the attributes displayed by the fans and supporters and ordinary Singaporean responses under each of the PERMA indicators.

Positive emotions

Positive emotions were quite evident especially during the hallmark event. Emotions such as pride, happiness and joy were well identified in the comments. Among the other social media, YouTube seems to capture the experience through the subscriber's comments almost immediately on the day of the post. This is best described in the following comments:

> All the goosebumps felt when the national anthem was played for the first time . . . couldn't control it. it's when your heart is so full of pride for Schooling, for his achievements, his dedication, for this tiny island nation,

> where finally, finally, changes are made to the sporting scene, it feels like my heart would burst. what a time to be alive.

> The part about our national anthem being played at the Olympics for the first time. I got goosebumps watching that live on tv, it felt surreal. It's probably the last thing we'd expect to happen before Joseph made history. Thank you Joe! Good luck for Tokyo 2020!

Even after the event, comments continued to stream in under the Olympic YouTube webpages. However, it was not difficult to single out the posts from Singaporeans. The video of the winning moment enabled the viewers to relive joy that the victory in August 2016 brought and continue to inspire those who watched it.

> This is like my 20th time watching it! Still get excited even though i know the final outcome. My heart just swelled with pride and so happy for Jo. His achievement transcends more than just swimming but about overcoming every obstacle to reach one's Goal!! Thanks Jo and family!

> Proud of you guys, you make ASEAN proud

After the event when Schooling returned to Singapore, the nation turned out in huge numbers for his victory parade where a large crowd waited for hours for his procession bus. By waving to him and shouting congratulations, the public shared their heartfelt joy and gratitude for the one who made them proud. He even had an orchid named in his honour – the *Dendrobium Joseph Schooling*. People still visit the YouTube video site to view Schooling's interview, recorded post-Olympic win and leave congratulatory comments, though the number of posted comments revealing positive emotions has tapered off over the last two years.

> proudest moment for every Singaporean. i still have tears in my eyes as I watch this.:')

> that was the moment i was crying CONGRATULATIONS TO Joseph Schooling

Engagement

The spectators may be watching the games away from the host city, but a strong connection was forged when they gathered at home, schools, work and a variety of social settings with an anticipation of the most coveted medal. The "flow" (Csikszentmihalyi, 1996) was very much visible from the way Singaporeans were fully absorbed by and focused on the event, starring Schooling.

> a proud Singaporean here! i still remember that time when literally EVERYONE in the country stopped to watch him on the tv. i was in a competition

> during that time though, and after that my teammates were all crowding around my teacher and she was showing us that he had won gold. ah the memories!:)
>
> As a Singaporean, I can remember very clearly the morning Joseph Schooling took the podium at the Olympics finals. My whole family was up, all my neighbours were up, and when he came in first, I could hear everyone around me cheering. It was such a surreal moment for the country, and I have never felt more proud to be Singaporean than when I heard our national anthem being played for the first time in history.

YouTube videos by Lim (2016) and Raymond (2016) depict the national solidarity, charged-up atmosphere and high levels of engagement. Accompanying this win was a celebration accorded to a celebrity at the airport and later the streets where he was paraded on an open decked bus (see Todayonline, 2016). The crowds waited for hours to have one glimpse of the national hero whose performance has kept them mesmerised. When analysing the data for post-event engagement, we did not find any further evidence of engagement. Perhaps, the recall of the "flow" experience indicating engagement does not sustain itself as much as other elements of the PERMA framework.

Relationships

The relationship indicator of PERMA is activated primarily because of an immediate relationship and connection the nation felt for their national sports ambassador. The home spectators became a part of his family, a fan, a follower and even the "worshippers". A national model of heroic proportions in sports was born that Singaporeans would brag about on their social media for many days to come.

The connection to the hallmark event for Singaporeans and the icon becomes more telling when thousands of expressions all over the mass media and social media picks up comments that are overwhelmingly positive and showcases a sense of belonging to a national project – winning the Olympics, at least 1 gold medal.

> When Joseph touched the wall and won, the whole estate screamed and made noises 😁 Someone even blasted the horn you would hear on Football matches😁😁😁
>
> I met his Father his parents live in the same estate as my grandparents
>
> I met this man at the airport after the NCAA champs last week. Took a pic with me even though he was with a friend. Truly an inspiration for me.

After the Rio Olympics, there appeared to be a "commodification" of the Olympic dream, which exploited the relationship built up between Schooling and his fans. Fans and supporters alike were interested in the lifestyle of their national hero and role model. Schooling went on to start a swim school and engaged himself in public events such as giving motivational talks to students in schools. Schooling also carries his own line of skincare products. Fashion and beverage brands such as Boss and Milo also caught up with him for product endorsements.

Meaning

Different meanings were experienced by individuals which are also reflected by the other indicators of PERMA. The Olympic Games has taken on a new meaning and hope for the nation with the first Olympic gold medal. The victory was shared by the entire nation. Here are the selected comments:

> His achievement transcends more than just swimming but about overcoming every obstacle to reach one's Goal!! Thanks Jo and family!

> Joseph I just loved the line: everyone is watching. The world is watching. Thank you so much for inspiring me becuz I'm going through the same phase of coming out of failure. Thank you so much for giving me hopes. I'll one day prove my myself. And yeah All the best. Wish to watch you more.

> Maybe it was perhaps he won my home country's first gold medal, but this moment really helped me escape the depression I was suffering at the time...

These quotes and others that were analysed, showed meaningfulness, which included hope, confidence, determination, resilience, role model, setting new goals, taming depression, perseverance, and the change of perspective among parents about sports education, career in sports and the accompanying benefits. These increasing gravitations and changed perspectives towards sports as a career were felt post-Olympic and sustained through local school's extracurricular activities and the enrolment in the Singapore Sports School (Tang, 2016). The viewers continue to leave comments on the social media sites, even after many years have passed. His victory continues to inspire and motivate those who look to him as a role model:

> I watched this over and over again, especially when i need a dose of inspiration. Singapore's golden boy. Jo Schooling

Accomplishment

To many fans and supporters, that event reminded them of accomplishment or Schooling is equivalent to the word "Accomplishment". Although they may not be Schooling, they tasted success through him. Imagine, following the Olympics time after time, but never the opportunity of hearing your own national anthem ever played? Now picture that moment when the flag is raised, and the anthem is played. Accomplishment along with all the other PERMA indicators such as positive emotions, meaning, relationships, also came into play as evident in these comments:

> all the goosebumps felt when the national anthem was played for the first time . . . couldn't control it. it's when your heart is so full of pride for Schooling, for his achievements, his dedication, for this tiny island nation, where finally, finally, changes are made to the sporting scene, it feels like my heart would burst. what a time to be alive.

> u just inspired many kids 2 join the sport just as @MichaelPhelps inspired u. This's incredible & in true spirit of Olympics.

> "You don't have to be a lawyer, a doctor or a successful businessman; there's success outside of those medium". More parents and people need to realise that.

A general spike in the number of swimming schools in Singapore resulted from parents' and general public's interest in the winning event "swimming". According to Tang (2016):

> Swim schools are seeing a spike in enquiries from potential students since Schooling's win. And lessons in his signature stroke – the butterfly – are turning out to be the hottest commodity.

Interest in enrolling their children in swimming classes perhaps enhanced Singaporeans' self-esteem. To be associated with an activity that brought laurels to the country gave them an opportunity to bask in the reflected glory, such that Schooling's achievement became their own accomplishment. Schooling's accomplishment at the event created a new sense of belonging for his fellow Singaporeans and these meanings will transcend through times and generations following the hallmark event.

Identity

The analysis of the text revealed a component other than the five from the PERMA framework. The findings revealed "identity" as an additional indicator of well-being and one that remains with the legacy that the hallmark event brought. With the accomplishment at the hallmark event comes the formation of a new national identity. Schooling became a talking point for the residents of the non-host city and the Eurasian ethnic community to which Schooling belongs. Both fans and supporters were quick to acknowledge the new chapter of history in the following comments.

> Never thought I live long enough to see a Singapore Eurasian win an Olympic gold medal for Singapore. Congratulations to you, your parents and Singapore.

Figure 5.2 Word cloud showing the visual analysis of textual data

> A true blue Singaporean 👍 even his dad has the Singlish accent

The word cloud in Figure 5.2 indicates a sense of "pride" in being a "Singapore" national and the "inspiration" from the swimming Olympics medal "win" appears to have a positive effect. "Singaporeans" were "amazed" and "inspired".

Conclusion

We started by highlighting the importance of hallmark events on a country's consciousness. Unless there is a continuum in maintaining the momentum driven from the first big event, the initial euphoria may dissipate over time. Understanding the context of the meaning of winning at least one Gold Medal in the Olympics for a tiny country, bereft of sporting talent given the small population compared to larger countries such as the USA and Russia, is a very big deal. The emotional aspect and the antecedent win only in regional games, meant that the Olympics gold medal was something euphoric that resonates with the Singapore citizenry.

Meaning, accomplishments and identity are three indicators that sustained even four years after the Rio Olympics. Although the other indicators were also present, they seem rather temporal unless one engages with the event and relives it again, for example, by watching the YouTube videos, or by being a part of the community that continues to follow Schooling's social media or the Olympic official page. After the event, the engagement between the community and the event is sustained through the hero, role model and ambassador of the hallmark event through other peripheral events that are related to them. The analysis revealed "identity" as an additional element other than the ones in the PERMA framework. The hallmark event encouraged people to talk about their cultural/national identities, feel proud of the shared identity with Schooling. However, identity did not appear to emerge as an independent component, rather seemed embedded within the five key PERMA components, inspiring us to propose a revised PERMAi model of well-being. This model hypothesises that one's well-being is also associated with their cultural/national identity. But a longitudinal study would be necessary to assess the next wave of PERMAi at the next Olympic Games in Tokyo to evaluate this proposed model.

Positive psychology theories and the PERMA model made a positive contribution to this scholarship on hallmark event, just as Filep, Volic and Lee (2015) suggest that appraising event experiences will help in future event planning and design. Further, this study has demonstrated that policymakers of non-host city of hallmark events such as the Olympics can invest and support their official participants in the event knowing that the social impact of national well-being extends beyond the event. The Rio Olympics left a legacy that the media has coined "the Schooling effect" (Ng, 2016; Fish, 2020). The new phenomenon saw parents deriving new meaning for the sport as analysed under the meaning indicator. It sparked new hope for athletes who were once told to abandon the Olympic dream (Ng, 2016). While this may all bring about positive outcome according to the PERMA analysis, we agree with Mezirow (1997) that the transformation plan needs to be part of the national objectives of the government to effect and support more changes.

While some scholars may advocate researchers to be site-specific in the selection of online sources for the selection of data, this study suggests that different social media sites may sustain different timelines of lived experiences. In the context of this study, we were able to analyse the well-being indicators using Twitter up till one month after the event. There appears to be no further posting about the games. The communication continues, however, on Facebook under Schooling's webpage. Interestingly, the indicators on PERMA were observable on YouTube up till the present time. Contrary to Whalen (2018) and Boellstorff *et al.* (2012), who argue that the unity of analysis should focus on a distinct community and not a channel based on standard reviews, we suggest future researchers using netnography to broaden the unit of analysis to also consider other websites when the data for the dedicated community's online media lacks information to provide meaningful analysis. Researchers need to be mindful of the social media culture in the local context. For example, Singaporeans may be more prone to using Facebook than Twitter – it is, therefore, important for researchers to understand which social media would contribute richer data and insights.

The significant contribution to hallmark event literature is the paradigm shift in defining or limiting studies to the host country. The impact, well-being of a community, nation, should all be part of the impact assessments – even for the

Olympics – not just the host cities and their residents but also the competing participants' and the homebound participants and fans watching from home. In a globalised world, as this case study shows, the audience or tourists at the event is just one segment of the participants. Because of global connectivity and the virtual world, audience from distant countries with ties to the event or interested in the event may find meaning in the outcomes. Hence, this research calls for more exploration and research that looks at the impacts of hallmark events beyond the immediate host community and event place.

References

Ascenso, S., Perkins, R., & Williamon, A. (2018). Resounding meaning: A PERMA well-being profile of classical musicians. *Frontiers in Psychology*. 9, 1–14.

Bakker, A.B., Schaufeli, W.B., Leiter, M.P., & Taris, T.W. (2008). Work engagement: An emerging concept in occupational health psychology. *Work Stress*. 22, 187–200.

Berry, L.L. (2002). Relationship marketing of services perspectives from 1983 and 2000. *Journal of Relationship Marketing*. 1 (1), 59–77.

Boellstorff, T., Nardi, B., Pearce, C., & Taylor, T.L. (2012). *Ethnography and virtual worlds: A handbook of method*. Princeton, NJ: Princeton University Press.

Branscombe, N.R., & Wann, D.L. (1991). The positive social and self concept consequences of sports team identification. *Journal of Sport and Social Issues*. 15 (2), 115–127.

Chirieleison, C., Montrone, A., & Scrucca, L. (2020). Event sustainability and sustainable transportation: A positive reciprocal influence. *Journal of Sustainable Tourism*. 28 (2), 240–262.

Choi, H., Kim, S., & Kwon, W. (2011). The causal relationship among the spectator motives, psychological well-being and behavioral intention of the spectators of international sports events [In Korean]. *Journal of Sport and Leisure Studies*. 46, 519–530.

Coghlan, A., & Filo, K. (2013). Using the constant comparison method and qualitative data to understand participants' experiences at the nexus of tourism, sport and charity events. *Tourism Management*. 35, 122–131.

Csikszentmihalyi, M. (1996). *Finding flow: The psychology of engagement with everyday life*. New York: Basic Books.

Csikszentmihalyi, M. (2013). *Flow: The psychology of happiness*. Sydney: Random House Australia.

Department of Statistics Singapore. (2020). *Population and population structure*. Singapore: Department of Statistics Singapore. Available from www.singstat.gov.sg/find-data/search-by-theme/population/population-and-population-structure/latest-data [Accessed 7 July 2020].

Downward, P., & Rasciute, S. (2011). Does sport make you happy? An analysis of the well-being derived from sports participation. *International Review of Applied Economics*. 25 (3), 331–348.

Doyle, J.P., Filo, K., Lock, D., Funk, D.C., & McDonald, H. (2016). Exploring PERMA in spectator sport: Applying positive psychology to examine the individual-level benefits of sport consumption. *Sport Management Review*. 19 (5), 506–519.

Edwards, D.J., & Steyn, B.J.M. (2008). Sport psychological skills training and psychological well-being. *South African Journal for Research in Sport, Physical Education and Recreation*. 30, 15–28.

Ericsson, K.A. (2002). Attaining excellence through deliberate practice: Insights from the study of expert performance. In: Ferrari, M. (ed.) *The pursuit of excellence through education*. Mahwah, NJ: Lawrence Erlbaum Associates, pp. 21–55.

Filep, S., Volic, I., & Lee, I.S. (2015). On positive psychology of events. *Event Management*. 19 (4), 495–507.

Filo, K., & Coghlan, A. (2016). Exploring the positive psychology domains of well-being activated through charity sport event experiences. *Event Management*. 20 (2), 181–199.

Fish, A. (2020, February 26). The 'schooling effect'. Available from https://swimlikeafish.org/the-schooling-effect [Accessed 7 June 2020].

Forgeard, M.J.C., Jayawickreme, E., Kern, M.L., & Seligman, M.E.P. (2011). Doing the right thing: Measuring wellbeing for public policy. *International Journal of Wellbeing*. 1 (1). DOI: 10.5502/ijw.v1i1.15.

Funk, D.C., & James, J.D. (2006). Consumer loyalty: The meaning of attachment in the development of sport team allegiance. *Journal of Sport Management*. 20, 189–217.

Getz, D. (2005). *Event management and event tourism* (2nd ed.). New York: Cognizant.

Grove, S.J., Dorsch, M.J., & Hopkins, C.D. (2012). Assessing the longitudinal robustness of spectators' perceptions of the functions of sport: Implications for sport marketers. *The Journal of Marketing Theory and Practice*. 20, 23–38.

Hall, C.M. (1987). The effects of hallmark events on cities. *Journal of Travel Research*. 26 (2), 44–45.

Hall, C.M. (1989). The definition and analysis of hallmark tourist events. *GeoJournal*. 19, 263–268.

Hall, C.M. (1996). Hallmark events and urban reimaging strategies: Coercion, community, and the Sydney 2000 Olympics. In: Harrison, L., & Husbands, W. (eds.) *Practicing responsible tourism: International case studies in tourism planning, policy, and development*. New York: John Wiley, pp. 366–379.

Hallmann, K., Breuer, C., & Kühnreich, B. (2013). Happiness, pride and elite sporting success: What population segments gain most from national athletic achievements? *Sport Management Review*. 16 (2), 226–235.

Kahneman, D., Diener, E., & Schwarz, N. (1999). *Well-being: The foundations of hedonic psychology*. New York: Russell Sage Foundation.

Kanste, O. (2011). Work engagement, work commitment and their association with well-being in health care. *Scandinavian Journal of Caring Sciences*. 25, 754–761.

Kavetsos, G., & Szymanski, S. (2010). National well-being and international sports events. *Journal of Economic Psychology*. 31 (2), 158–171.

Kozinets, R.V. (2002). The field behind the screen: Using netnography for marketing research in online communities. *Journal of Marketing Research*. 39 (1), 61–72.

Kozinets, R.V. (2015). *Netnography: Redefined* (2nd ed.). London: Sage.

Kozinets, R.V., Dolbec, P.-Y., & Earley, A. (2014). Netnographic analysis: Understanding culture through social media data. In: Flick, U. (ed.) *Sage handbook of qualitative data analysis*. London: Sage, pp. 262–275.

Lai, M.K., Leung, C., Kwok, S.Y.C., Hui, A.N.N., Lo, H.H.M., Leung, J.T.Y., & Tam, C.H.L. (2018). A multidimensional perma-h positive education model, general satisfaction of school life, and character strengths use in Hong Kong senior primary school students: Confirmatory factor analysis and path analysis using the APASO-II. *Frontiers in Psychology*. 9, 1–11.

Lee Duckworth, A., Steen, T.A., & Seligman, M.E. (2005). Positive psychology in clinical practice. *Annual Review of Clinical Psychology*. 1, 629–651.

Lim, D. (2016, September 2). Joseph schooling Olympic 2016 winning swim [video]. Available from https://youtu.be/TcRgHFoZUTQ [Accessed 7 July 2020].

Linley, P., Joseph, S., Harrington, S., & Wood, A. (2006). Positive psychology: Past, present, and (Possible) future. *The Journal of Positive Psychology*. 1, 3–16.

Mahony, D.F., Madrigal, R., & Howard, D.R. (1999). The effect of individual levels of self-monitoring on loyalty to professional football teams. *International Journal of Sports Marketing & Sponsorship*. 1, 146–167.

Mezirow, J. (1997). Transformative learning: Theory to practice. *New Directions for Adult and Continuing Education*. 74, 5–12.

Morgan, R.M., & Hunt, S.D. (1994). The commitment-trust theory of relationship marketing. *Journal of Marketing*. 58, 20–38.

Ng, J. (2016, August 21). The schooling effect. *The New Paper*. Available from www.tnp.sg/news/singapore-news/schooling-effect [Accessed 7 July 2020].

Nind, M., Wiles, R., Bengry-Howell, A., & Crow, G. (2012). Methodological innovation and research ethics: Forces in tension or forces in harmony? *Qualitative Research*. 13 (6), 650–667.

Ouyang, Z., Gursoy, D., & Chen, K. (2019). It's all about life: Exploring the role of residents' quality of life perceptions on attitudes toward a recurring hallmark event over time. *Tourism Management*. 75, 99–111.

Powlowski, T., Downward, P., & Rasciute, S. (2014). Does national pride from international sporting success contribute to well-being. *An International Investigation. Sport Management Review*. 17, 121–132.

Rasciute, S., & Downward, P. (2010). Health or happiness? What is the impact of physical activity on the individual. *Kyklos*. 63 (2), 256–270.

Raymond. [Raymond's Math & Science Studio]. (2016, August 14). School's golden moment telecast live at our school at Bishan [video]. Available from https://youtu.be/2ipS-8k4FPE [Accessed 7 June 2020].

Ritchie, B.J.R. (1984). Assessing the impact of hallmark events: Conceptual and research issues. *Journal of Travel Research*. 23 (1), 2–11.

Ryan, R.M., & Deci, E.L. (2000). Self-determination theory and the facilitation of intrinsic motivation, social development, and well-being. *American Psychologist*. 55, 68–78.

Ryan, R.M., & Deci, E.L. (2001). On happiness and human potentials: A review of research on hedonic and eudaimonic well-being. *Annual Review of Psychology*. 52, 141–166.

Ryan, R.M., Sheldon, K.M., Kasser, T., Deci, E.L., Gollwitzer, P.M., & Bargh, J.A. (eds.) (1996). *The psychology of action: Linking cognition and motivation to behavior*. New York: Guilford Press.

Ryff, C.D. (1989). Happiness is everything, or is it? Explorations on the meaning of psychological well-being. *Journal of Personality and Social Psychology*. 57, 1069–1081.

Ryff, C.D., & Keyes, C.L.M. (1995). The structure of psychological well-being revisited. *Journal of Personality and Social Psychology*. 69, 719–727.

Seligman, M.E.P. (2012). *Flourish: A visionary new understanding of happiness and well-being*. New York: Free Press.

Seligman, M.E.P., & Csikszentmihalyi, M. (2000). Positive psychology: An introduction. *American Psychologist*. 55 (1), 5–14.

Tang, A. (2016, August 18). Swim school gain from 'butterfly effect'. The Straits Times. Available from www.straitstimes.com/singapore/swim-schools-gain-from-butterfly-effect [Accessed 5 July 2020].

Todayonline. (2016, August 15–19). Joseph Schooling whirlwind homecoming [video]. Available from https://youtu.be/n6Ryef_cjjU [Accessed 7 June 2020].

Todd, L., Leask, A., & Ensor, J. (2017). Understanding primary stakeholders' multiple roles in hallmark event tourism management. *Tourism Management*. 59, 494–509.

Trail, G., Anderson, D.F., & Fink, J. (2000). A theoretical model of sport spectator consumption behavior. *International Journal of Sport Management*. 1, 154–180.

van der Steen, T., & Richards, G. (2019). Factors affecting resident support for a hallmark cultural event: The 2018 European capital of culture in Valletta, Malta. *Journal of Policy Research in Tourism, Leisure and Events*. DOI: 10.1080/19407963.2019.1696352.

van Hilvoorde, I., Elling, A., & Stokvis, R. (2010). How to influence national pride? The Olympic medal index as a unifying narrative. *International Review for the Sociology of Sport*. 45 (1), 87–102.

Viol, M., Todd, L., Theodoraki, E., & Anastasiadou, C. (2018). The role of iconic-historic commemorative events in event tourism: Insights from the 20th and 25th anniversaries of the fall of the berlin wall. *Tourism Management*. 69, 246–262.

Waitt, G. (2003). Social impacts of the Sydney Olympics. *Annals of Tourism Research*. 30 (1), 194–215.

Wang, R.T., Ho, C.M., & Zhang, J.J. (2012). Examining the effects of relationship quality and calculative commitment on sport consumer behaviors for intercollegiate athletics. *Journal of Issues in Intercollegiate Athletics*. 5, 301–328.

Wang, T.R., Min, S.D., & Kim, S.K. (2013). Fulfillment of sport spectator motives: The mediation effect of well-being. *Social Behavior and Personality*. 41 (9), 1421–1433.

Wann, D.L. (2006). Understanding the positive social psychological benefits of sport team identification: The team identification-social psychological health model. *Group Dynamics: Theory, Research, and Practice*. 10 (4), 272–296.

Wann, D.L., Hackathom, J., & Sherman, M.R. (2017). Testing the team identification social psychological health model: Mediational relationships among team identification, sport fandom, sense of belonging and meaning in life. *Group Dynamics: Theory, Research, and Practice*. 21, 94–107.

Wann, D.L., Inman, S., Ensor, C.L., Gates, R.D., & Caldwell, D.S. (1999). Assessing the psychological well-being of sport fans using the profile of mood states: The importance of team identification. *International Sports Journal*. 3, 81–90.

Wann, D.L., & James, J.D. (2019). *Sport fans: The psychology and social impact of fandom* (2nd ed.). New York: Routledge.

Wann, D.L., & Weaver, S. (2009). Understanding the relationship between sport team identification and dimensions of social well-being. *North American Journal of Psychology*. 11, 219–230.

Whalen, E.A. (2018). Understanding a shifting methodology: A content analysis of the use of netnography in hospitality and tourism research. *International Journal of Contemporary Hospitality Management*. 30 (11), 3423–3441.

Williams, J.M. (2009). *Applied sport psychology: Personal growth to peak performance* (6th ed.). London: McGraw-Hill Higher Education.

6 Advice for Advent in Zagreb organisers

Insight into well-being and quality of life

Ela Brolich, Marko Perić and Nicholas Wise

Introduction

Tourism is one of the largest industries in the world, with popular events at its core. People travel to participate in a variety of sports, business or cultural events and annual happenings. These are important for destinations because they receive a boost in tourism with various events. In general, the impact that events have on a destination is evident in economic, socio-cultural and environmental dimensions (Richards, de Brito & Wilks, 2013; Wise, 2016; Gursoy *et al.*, 2019). Within each of these dimensions, it is possible to identify the positive and negative effects resulting from the way individual events, and a destination as a whole, are managed.

Tourism developments can improve destination image, extended tourist season, increase tourism-generated revenue and jobs, but occasionally, higher prices and costs of living in a destination can have a direct impact on residents and local economic well-being in a destination (Spirou, 2011; Richards, 2017; Kersulić, Perić & Wise, 2020). Socio-cultural effects primarily include interaction among people, new knowledge acquisition, opportunities to learn about other cultures (i.e. the cultures of tourists), greater cultural tolerance, and can increase a sense of place and community pride among residents, but also undesirable effects such as more crime and vandalism (Colombo, 2016; Pavluković, Armenski & Alcántara-Pilar, 2017; Wise & Harris, 2017; Perić, 2018). All large gatherings of people carry a greater burden on the environment and municipal structure from more waste, heavier traffic and increased noise or light pollution. Alternatively, however, if events are designed in a planned and methodical way, they can help to raise environmental awareness and regenerate abandoned or neglected regions (Smith, 2012). All stakeholders – residents as well as businesses and destination managers and planners – in a destination feel these effects.

Recently, more and more focus is being placed on the impact of tourism on the quality of life of residents, who are recognised as a major stakeholder in the strategic planning of tourism development in any destination (Đurkin & Perić, 2017; Đurkin & Wise, 2018). Residents directly experience both the positive and the negative economic, socio-cultural and environmental effects of every event and of tourism in general, making it necessary to continuously monitor the impact of tourism on residents. On the whole, holding an event does not necessarily need to have any adverse outcomes for residents (Zhao & Wise, 2019). Likewise, understanding and adjusting to the needs of residents will not negatively impact on an event or tourism activities, because residents will also be among the people buying and consuming the products and services offered during an event. Moreover, each of the aforementioned effects can be linked to the residents' quality of life and overall well-being. Thus, when planning events, event organisers need to pay special attention to the quality of life and well-being of residents.

Quality of life and well-being are complex, multi-layered and intangible phenomena and, as terms, are often used in synergy or interchangeably (Iwasaki, 2007; Uysal *et al.*, 2016; Perles-Ribes *et al.*, 2020;). Quality of life is the level of well-being that individuals feel – and, hence, subjective well-being is a key determinant of quality of life (Kim, Uysal & Sirgy, 2012) and the overall liveability in a place that experiences high concentrations of visitors (see Nofre *et al.*, 2018). In addition to well-being, quality of life also includes happiness and life satisfaction (Dagger & Sweeney, 2006; Uysal *et al.*, 2016; Moon *et al.*, 2019; Ouyang, Gursoy & Chen, 2019). Accordingly, both quality of life and well-being possess objective and subjective dimensions. The objective dimensions of quality of life (or well-being) primarily refer to the measurable economic, environmental or health-related aspects of living, and the subjective dimensions, to specific psychological constructs such as happiness, satisfaction and perceived quality of life (Uysal *et al.*, 2016; Bimonte, D'Agostino & Grilli, 2020).

The above suggests that people's quality of life is a set of measurable and non-measurable factors open to the subjectivity of interpretation. In a certain sense, "quality of life" is an umbrella term (Yolal *et al.*, 2016) and, as such, is the subject of study of many researchers, in particular, in the context of tourism and the effects that tourism development has on the quality of life of tourists and residents alike (Moscardo, 2009; Carneiro & Eusebio, 2015; Musikanski *et al.*, 2019; Bimonte, D'Agostino & Grilli, 2020). Previous studies tend to also suggest that the economic benefits of tourism (e.g. increased revenues or new jobs) positively affect quality of life and well-being. However, certain economic (e.g. higher prices), social (increased crime level) and ecological (e.g. increased waste and noise level) issues are linked to lower levels of residents' satisfaction (e.g. Kaplanidou, 2012; Liang & Hui, 2016; Ouyang, Gursoy & Chen, 2019; Perles-Ribes *et al.*, 2020). For this reason, quality of life is often referred to as a mediating variable between the effects of a given tourism event and the support of residents for that event and for tourism in general (Kaplanidou *et al.*, 2013; Ma & Kaplanidou, 2016; Ouyang, Gursoy & Chen, 2019).

The evaluation of effects and residents' quality of life have been studied in a variety of different contexts but primarily in the context of sports and cultural events. When dealing with cultural events, research often focuses on music festivals and large-scale events such as the European Capital of Culture (ECoC) project (Han *et al.*, 2017; Pavluković, Armenski & Alcántara-Pilar, 2017). While recognised as a global phenomenon and a generator of economic effects (Marcher, Erschbamer & Pechlaner, 2019), Christmas markets have rarely been the subject of scientific papers. Brida, Disegna and Osti (2013) and Brida, Meleddu & Tokarchuk (2017) analysed tourist consumption, or the economic and social value of tourists (the so-called consumer surplus), at Christmas markets in northern Italy but did not take residents into consideration. With regard to Christmas markets and their effects on residents, these markets are seen as a powerful tool for boosting a sense of community and local identity (Winchester & Rofe, 2005) but can also lead to excessive commercialisation and over-crowding, which residents experience as extremely negative effects (Marcher, Erschbamer & Pechlaner, 2019). As these studies focus on analysing the individual effects of Christmas markets, only indirectly do they examine the quality of life of residents.

Hence, this chapter aims to explore the effects of the Advent in Zagreb (Croatia) event on residents' quality of life and to identify which of the event's elements can be improved. The purpose of this research is to find ways to maximise positive experiences at the event for residents as well as tourists. The following section establishes the context of the research. Next, the methodological approach is described and results are presented. The chapter ends with the discussion and conclusion section.

Case study – Advent in Zagreb, Croatia

Zagreb is the capital city of the Republic of Croatia. It is the country's main educational, commercial and cultural hub and home to about 800,000 inhabitants. Until some ten years ago, Zagreb was, in terms of tourism, just an incidental stop for tourists on their way to the traditionally highly developed coastal destinations. Today, thanks to its abundant cultural facilities and services, diverse events, and multitude of restaurants, shopping centres, and leisure and entertainment venues, Zagreb has become a destination on an equal footing with seaside destinations and is seeing growth in tourism traffic year-after-year. In 2019, Zagreb recorded about 1.4 million tourist arrivals and 2.5 million tourist overnights, accounting for 7.4% of the total number of tourist arrivals and only 2.9% of total tourist overnights in Croatia. Zagreb has a diverse accommodation offering, made up of numerous hotels in a wide range of categories as well as hostels that are popular among younger tourists.

Held annually since 2010, Advent in Zagreb is the city's most important event. Zagreb city officials with the help of the Zagreb City Tourist Board organise the event and focus efforts on ensuring the quality and availability of information concerning the event is widespread. Advent in Zagreb is a traditional event organised every year and has a strongly pronounced cultural component throughout nurturing not only Christian traditions but also local cuisine and customs. Each year, the official lighting of Advent candles in Zagreb begins four weeks before Christmas, with the lighting of the first candle on the Advent wreath. Over the years, the event's offering has visibly expanded and its duration has been extended not only to satisfy the wants and needs of visitors but also to help maximise the revenue generated. A challenge organisers need to pay attention is not to over-commercialise the event as this can downgrade Zagreb's differentiation from close competition, and a focus on increasing profits might downplay the shared sense of community that the event highlights through its Christian motives. Up to 2019, the tourism offering of Advent in Zagreb was largely concentrated in Zagreb's Upper Town and Lower Town, that is, the city centre. Since then, the event has spread to other parts of the city due to very high demand and to help reduce crowds in the downtown area. As indicated, as the event expands, it is important to focus on community ideals opposed to liberalising business practices.

The most important attractions in the Upper Town are the live nativity scene, especially appealing to children because it allows them to interact with domestic animals; Advent on Stross (i.e. on the Strossmayer Promenade), which many visitors find magical, in particular, because of the live concerts; and Advent at Klovićevi Dvori and the Gradec Plateau. Highlighted attractions in the Lower Town are Advents on: Ban Jelačić Square, European Square in the vicinity of Ban Jelačić Square, in Grič Tunnel (with its highly appealing decorations), and at Zrinjevac (with numerous concerts held in a storybook setting). Advent Fooling Around is an event which hosts musicians and exhibitors in a number of different venues since 2011. The Ice Park, on King Tomislav Square, is also a popular place for the events gastronomic offering since 2014. In addition to its entertainment offering, Advent in Zagreb boasts a rich international as well as traditional local culinary offerings. Figures 6.1, 6.2 and 6.3 show some of the settings, attractions and events and how the Advent in Zagreb promotes social gathering.

Figure 6.1 Advent atmosphere in Zagreb

Source: First author.

Figure 6.2 Ban Josip Jelačić Square during Advent

Source: First author.

Figure 6.3 Programme and Advent decoration

Source: First author.

The strategic management and planned development of Advent in Zagreb over the past decade have earned Zagreb the title of Best Christmas Market – recognising it as the best European destination to visit during Advent – three years running, in 2016, 2017 and 2018. Having won the title three years in a row, Zagreb was no longer eligible for nomination for the title in 2019. Although Zagreb is not the only Croatian city to annually organise this type of event, for many people, impression of Zagreb immediately brings to mind Advent – suggesting that Zagreb's Advent is associated with city branding. Given that Zagreb was not eligible for a nomination in 2019 and 2020 will likely be impacted due to COVID-19, it will be important for city planners to decorate the city during the time and broadcast the event virtually so that they can reinforce this popular imagination of the city, and this can potentially extend the impact of the event's image even further to attract visitors in future years for a live experience.

Methodology

To explore the impact of Advent in Zagreb on the quality of life of the city's residents, empirical research involving residents was conducted. A questionnaire was designed consisting of three parts and written in the Croatian language. The first part comprises a total of four questions, of which three are structured dichotomous questions and one, a multiple-choice question. The three closed-ended questions, allowing only one answer, focus on the frequency of visiting the event and the mode of transport to the event venues, while the multiple-choice question refers to the reasons for attending the event. The second part of the questionnaire consists of 20 statements referring to the event's impact on Zagreb, the assessment of the offering's availability and diversity, the accessibility of the offering in terms of price, satisfaction with the offering and the way it is organised, and satisfaction with the way traffic is managed during the event. The statements were modified and context-adjusted after Kim *et al.* (2015) and Perić (2018). The respondents were also asked to express agreement/disagreement with statements about the event having a good effect on their, or other residents, quality of life. Both statements were formulated after Kaplanidou *et al.* (2013), Ma and Kaplanidou (2016) and Moon *et al.* (2019). Finally, respondents were asked whether they agreed/disagreed that they would recommend others to visit Advent in Zagreb. Their level of agreement with each statement was measured using a 5-point Likert scale (1 = Strongly disagree, 2 = Somewhat disagree, 3 = Neither agree nor disagree, 4 = Somewhat agree, 5 = Strongly agree). The third and last set of questions refers to demographic data and examines the respondents' age, gender, educational level and the length of time they have been living in Zagreb.

Data gathering was carried out at the end of December 2019 and the beginning of January 2020, on a convenience sample of Zagreb's inhabitants. This timeframe was set to give the respondents enough time to explore all the elements of Advent in Zagreb and to form their own opinions regarding the elements under research. In this time period, respondents were able to fill out the online questionnaire using Google Forms. To ensure that it reached diverse demographic groups, the questionnaire was also posted on the Facebook page Zakaj volim Zagreb (Why I love Zagreb), and most of the filled-in questionnaires were received as a result of this posting. The researchers collected 436 filled-in questionnaires, of which 419 were complete and usable for analysis.

The agreement scale scores were analysed using descriptive statistics, and bivariate analysis was used to determine the relationship between the perception of local residents about individual effects of the event and their personal quality of life.

Research results

Women were more inclined to participate in the survey, resulting in 333 women (80% of respondents) filling in the questionnaire and 86 men (20% of respondents) (Table 6.1). One in four respondents has secondary school qualifications. Most of the respondents (228 or 56.5%) hold a college or university degree, and most of them are between the ages of 26 and 45. Fully 71% of the respondents reported being born and raised in Zagreb. The question "How long have you been living in Zagreb?" was open-ended. Respondents born and raised in Zagreb gave responses such as "all my life", "since I was born" and "continuously, since I was born". The age group of respondents giving such responses was heterogeneous. The least number of respondents reported that they have been living in Zagreb for the past 15 to 20 years, meaning that in some previous period of their lives, most likely when they were young, they moved to Zagreb from other places. This result was expected, given the analysis of respondents by age groups. Namely, all respondents who reported they have been living in Zagreb for the past 15 to 20 years belong to the middle-aged or older age groups (least represented in this study).

Table 6.1 Analysis of respondents' demographic data

Gender	*N (419)*	*%*	*Educational Background*	*N (419)*	*%*
M	86	20.77%	Elementary school	0	0.00%
F	333	80.43%	Secondary school	108	26.09%
Age	***N* (419)**	**%**	College or university	234	56.52%
18–25	91	21.98%	Master's degree or postgraduate specialist studies	77	18.60%
26–35	128	30.92%	**Length of residence in Zagreb (in years)**	***N* (419)**	**%**
35–45	108	26.09%	Up to 5	40	9.66%
46–55	65	15.70%	Up to 10	27	6.52%
Over 56	26	6.28%	Up to 15	18	4.35%
			Up to 20	12	2.90%
			More than 20	35	8.45%
			Since birth	287	69.32%

Most of the respondents (81.4%) reported that prior to filling in the questionnaire they had attended at least one event held as part of Advent in Zagreb (Table 6.2). Most of the respondents paid several visits to such events, with 13.5% reporting they visited these events often. A total of 18.6% of respondents did not attend Advent in Zagreb. Such questions related to attending allowed for multiple responses. Hence, the scores in this question are higher than those in other questions. Table 6.2 presents the possibility that Advent in Zagreb itself was not the main motivation for residents to attend. The majority of respondents reported that they had enjoyed spending time with friends at the event, suggesting that their primary motivation was in fact social contact, regardless of the event. The event venues confirm this finding. Namely, the Advent in Zagreb venues are located in the city's most frequented locations. The true indicator of residents' interest in visiting the event is also evident in "enjoying the ambiance/atmosphere of the event", which is in line with social motives. "Cultural events" and "gastronomic offering" were less-mentioned reasons for attending the event. Other reasons for visiting were "daily movement" (e.g. to school or work), "curiosity" and "accompanying others" (children, friends visiting Zagreb, etc.).

The smallest share of respondents (14.6%) exclusively uses cars for getting to Advent in Zagreb venues. One of the response options to this question was "I use a combination of at least two of the response choices", and 30% of responses contained this response option, which does not rule out the use of cars as one of the two types of transport. Hence, it is possible that the respondents travelled to a venue by car and then continued either on foot or by public transport. This assertion could be true considering the very limited number of parking spaces in Zagreb's centre and because the use of a combination of transport modes is in fact a very frequent practice of residents all year round, not only during Advent in Zagreb.

Table 6.3 presents the results relating to various aspects of the offering and the effects that the event has on Zagreb and its inhabitants. The respondents mostly

Table 6.2 Analysis of visits by residents

Frequency of Visits	*N (419)*	*%*	*Reasons for Visiting**	*N*	*%*
I did not visit the event	77	18.60%	Gastronomy	84	20.29%
Once	237	57.25%	Atmosphere/ambience	242	58.45%
Several times	49	11.84%	Social interaction	261	63.04%
Often	56	13.53%	Cultural events	75	18.12%
Transport modes	***N* (363)**	**%**	Other reasons	25	6.04%
Car	53	14.60%			
Walking	49	13.50%			
Public transport	152	41.87%			
Combination of the above	109	30.03%			

*Multiple answers possible.

Table 6.3 Analysis of respondents' degree of agreement regarding the offering and impact of Advent in Zagreb

Statement	*1*	*2*	*3*	*4*	*5*	*Me*	*Mo*	*SD*
	%	%	%	%	%			
I think Advent in Zagreb has a diverse offering.	5.2	16	25.5	22	32.2	3.62	5	1.24
I think Advent in Zagreb has an easily accessible offering.	6.7	18.3	26	27.5	21.5	3.42	4	1.20
I think the prices of products match their quality.	31.7	30.7	28.5	7.2	2	2.20	2	1.03
The prices of food and beverages at Advent in Zagreb are too high.	4.2	6.4	19.6	19.6	50.2	4.01	5	1.17
The prices of products and services have increased during Advent in Zagreb.	8.4	13.6	27.5	25.8	25.2	3.41	3	1.26
The number of entertainment and cultural events has increased during Advent in Zagreb.	5.2	12.6	25	30.7	26.5	3.59	4	1.17
I think the Advent in Zagreb offering helps promote Zagreb's cultural heritage.	19.1	29	29.2	14.6	8.2	2.64	3	1.19
I think Advent in Zagreb helps encourage the preservation of local culture.	16.1	25.5	28.7	18.6	11.1	2.83	3	1.23
Traffic in downtown Zagreb is well organised during Advent in Zagreb.	20.3	22	24.5	20.8	12.4	2.84	3	1.32
I think free Zagreb Electric Tram (ZET) trams and buses on weekends help resolve traffic congestion during Advent in Zagreb.	16.3	15.8	26.2	17.1	24.5	3.13	3	1.40
I find annoying the crowds at Advent in Zagreb venues.	13.6	15.3	21	19.6	30.4	3.33	5	1.41
I find annoying the noise generated by various Advent in Zagreb events.	36.4	23.5	18.8	11.6	9.7	2.34	1	1.33
I think Advent in Zagreb has a negative impact on the environment.	27.2	21.5	27	13.1	11.1	2.57	1	1.30
I think there is risk of a terrorist attack in Zagreb during Advent in Zagreb.	44.6	26	19.6	7.2	2.7	1.96	1	1.10

Note: Me = Mean; Mo = Mode; *SD* = Standard Deviation (repeated in subsequent tables accordingly).

find the offering diverse and easily accessible (average scores are 3.6 and 3.4, respectively). An issue regarding Advent in Zagreb is that most of the respondents think the prices of products and services do not match their quality, that is, they consider the prices of products and services at the event to be too high. Namely, only 9.2% of respondents gave scores of 4 or 5 to confirm they agree that the prices of products and services are proportional to their quality but as many as 31.7% of respondents gave this statement a score of 1, meaning they strongly disagree. More than half of the respondents (50.2%) strongly agree with the statement that the prices of products and services at the event are too high. Moreover, they believe that prices are higher during the event than other times of the year.

Although the respondents mostly did not confirm that Advent in Zagreb encourages the preservation of the local culture, they largely agree with the statement that the number of cultural and entertainment events held during Advent in Zagreb has increased. Furthermore, the respondents were moderately negative with regard to traffic during Advent in Zagreb. On average, they see some differences between traffic during the event and traffic during the rest of the year. They are not satisfied with how the traffic is organised but are not sure whether free transport on Zagreb Electric Tram (ZET) trams and buses over the Advent weekends helps to resolve the issue of traffic congestion. Notably, however, half of the respondents (50%) stated that they are annoyed by event venue crowding. The respondents were less critical of the noise generated by the various events held during Advent in Zagreb. The average agreement scale score for this statement was 2.3 while the score of 1 – Strongly disagree – was convincingly the most chosen score, selected by 36% of respondents. The average agreement scale score of the statement "I think Advent in Zagreb has a negative impact on the environment" is 2.6, suggesting that residents are not concerned about the impact of the event on the environment. Of all the statements, the respondents least agreed with the statement "I think there is risk of a terrorist attack in Zagreb during Advent in Zagreb". This statement has an average agreement scale score of 2.0, with as many as 44.6% of respondents reporting that they strongly disagree. Only 9.9% of respondents reported (with a score of 4 or 5) they believe that the risk of an attack exists to a certain extent.

With regard to residents' satisfaction, they are not satisfied with the way traffic is organised in the city centre (2.8). Considering that most of the respondents did not have any bad experiences with the products and services offered at the event, the general satisfaction of respondents is slightly above the average (3.4) (Table 6.4).

The last three statements refer to perceived personal quality of life, residents' quality of life and making recommendations to others to visit the event (Table 6.5). Almost a half of the respondents (47.7%) gave a score of 1 or 2 to the statement that Advent in Zagreb has a positive effect on their personal quality of life, and almost a third of respondents (32.2%) could not assess whether Advent in Zagreb

Table 6.4 Analysis of respondents' degree of agreement with statements concerning satisfaction

Statement	*1*	*2*	*3*	*4*	*5*	*Me*	*Mo*	*SD*
	%	*%*	*%*	*%*	*%*			
I am satisfied with the way traffic is organised during Advent in Zagreb.	19.3	25	22.3	20.5	12.9	2.83	2	1.31
I have had bad experiences with the products and services offered at Advent in Zagreb (e.g. poor quality of food).	26.2	24.8	25.2	13.6	10.1	2.52	1	1.29
In general, I am satisfied with the Advent in Zagreb offering.	7.2	18.3	26	27.2	21.3	3.38	4	1.22

Table 6.5 Analysis of respondents' degree of agreement with statements concerning quality of life

Statement	*1*	*2*	*3*	*4*	*5*	*Me*	*Mo*	*SD*
	%	%	%	%	%			
Advent in Zagreb has a good effect on my quality of life.	26.7	21	26	15.8	10.4	2.65	1	1.33
Advent in Zagreb has a good effect on the quality of life of residents.	12.4	20.5	32.2	21	13.9	3.04	3	1.22
I would recommend others to visit Advent in Zagreb.	5.2	11.1	23.8	21	38.9	3.77	5	1.23

Table 6.6 Spearman's rank-order correlation coefficient, "rho", calculated between individual statements and perceived quality of life

Statement	*Advent in Zagreb Has a Good Effect on My Quality of Life*	
	Correlation Coefficient	*Sig.*
I think Advent in Zagreb has a diverse offering.	0.500	0.000
I think Advent in Zagreb has an easily accessible offering.	0.541	0.000
I think the prices of products match their quality.	0.507	0.000
The prices of food and beverages at Advent in Zagreb are too high.	−0.006	0.902
The prices of products and services have increased during Advent in Zagreb.	0.024	0.619
The number of entertainment and cultural events has increased during Advent in Zagreb.	0.553	0.000
I think the Advent in Zagreb offering helps promote Zagreb's cultural heritage.	0.498	0.000
I think Advent in Zagreb helps encourage the preservation of local culture.	0.533	0.000
Traffic in downtown Zagreb is well organised during Advent in Zagreb.	0.477	0.000
I think free Zagreb Electric Tram (ZET) trams and buses on weekends help resolve traffic congestion during Advent in Zagreb.	0.357	0.000
I find annoying the crowds at Advent in Zagreb venues.	−0.162	0.001
I find annoying the noise generated by various Advent in Zagreb events.	−0.200	0.000
I think Advent in Zagreb has a negative impact on the environment.	−0.315	0.000
I think there is risk of a terrorist attack in Zagreb during Advent in Zagreb.	−0.078	0.113
I am satisfied with the way traffic is organised during Advent in Zagreb.	0.461	0.000
I have had bad experiences with the products and services offered at Advent in Zagreb (e.g. poor quality of food).	−0.235	0.000
In general, I am satisfied with the Advent in Zagreb offering.	0.634	0.000
I would recommend others to visit Advent in Zagreb.	0.616	0.000

has a favourable impact on residents' quality of life. This suggests that while most of the respondents believe the event does not contribute to improving the quality of their personal lives, they were indecisive as to whether it helps to enhance the quality of life of residents as a whole. In other words, the respondents feel that the event has a more favourable effect on the quality of life of others (3.0) rather than on their own quality of life (2.7). In general, despite not being overly satisfied with the individual elements of the offering, prices, etc., the respondents would nevertheless recommend others to visit Advent in Zagreb (Table 6.5).

Finally, Spearman's rank-order correlation coefficient, "rho", was calculated between the residents' perceptions of the individual effects of the events and the residents' perceptions of how the event affects their personal quality of life (Table 6.6). It is evident that there is a moderate but statistically significant positive correlation between the diversity and accessibility of the offering, the price-quality match, the number of cultural and entertainment events and well-organised traffic on the one hand and the perceived quality of life of residents on the other hand. There is also a weak but statistically significant positive correlation between free public transport and perceived quality of life, and a weak but statistically significant negative correlation between generated noise, environmental impacts and bad experiences on the one hand and perceived quality of life on the other. In addition, there is a weak but statistically significant negative correlation between traffic congestion and quality of life. As expected, general satisfaction with the offering at Advent in Zagreb and the probability of recommending the event to others are positively correlated with perceived quality of life (with a moderate statistically significant correlation).

Discussion

The conducted study suggests that residents do not perceive Advent in Zagreb as a contributor to their quality of life. However, not all factors have the same influence. Respondents feel the offering at Advent in Zagreb is rather good and that the number of entertainment and cultural events increases, with the correlation between these variables and perceived quality of life is positive but moderate. It could be speculated that the diverse and easily accessible offering, inclusive of cultural events, has a favourable effect on residents' quality of life. On the other hand, most of the residents think prices have gone up, making them too high relative to product quality and the residents' purchasing power. Zhao and Wise (2019) made a similar finding, but respondents in their study were not discouraged by higher prices because they were supporting members of the community. Although respondents have not had negative experiences with product quality at the event in Zagreb, other studies have found that high prices can constrain residents from enjoying the products (Kim *et al.*, 2015; Kahraman & Kazançoğlu, 2019), resulting in satisfaction declines during the event. This study, however, did not confirm such a correlation between (too high) prices and quality of life.

Concerning other crucial elements affecting quality of life, it should be noted that respondents, in general, think the event does not contribute to the increase of environmental pollution, noise levels or risk of a terrorist attack. This finding leads to the conclusion that residents largely feel they are safe and in a healthy environment. While Carneiro and Eusebio (2015) reached a similar conclusion, some other studies contradict it and suggest a decline in residents' sense of safety (Bimonte, D'Agostino & Grilli, 2020). This study further indicates that traffic problems exist to a certain extent, but free public transport during weekends is not a perceived solution to under-capacity of parking places, which is a year-round problem. Hence, Advent in Zagreb constricts the daily functional movement of the city's inhabitants, increasing the residents' level of dissatisfaction. Overtourism – the increase in tourist arrivals over a short time, exceeding the carrying capacity of a destination – has been shown to have adverse effects on the quality of life of residents (see Musikanski *et al.*, 2019), but in this study such a correlation is weak. This relates to findings by Nofre *et al.* (2018), where residents see sharp increases in visitors disrupt local well-being and liveability in a place. While events aim at achieving economic impacts, we must not forget that the local disruptions that they cause directly impact well-being (Bimonte, D'Agostino & Grilli, 2020).

While residents are neutral on the effect Advent in Zagreb has on others' quality of life, this is perceived more favourably compared with how residents feel the event has increased their own quality of life. The reason for such a result could relate to differences in perceiving the subjective and objective elements of residents' quality of life (Yolal *et al.*, 2016; Bimonte, D'Agostino & Grilli, 2020). It is possible that Zagreb's inhabitants took into greater consideration the objective quality of life elements (such as measurable personal financial benefits) when asked to evaluate the effect of Advent in Zagreb on their personal quality of life, and the subjective elements (such as the quality of social interaction) concerning the effect of the event on the city's population. This result aligns with the work by Kim, Uysal and Sirgy (2012) that these perceptions of quality of life are based on different ways of interpreting subjective well-being, and how they make portray their own perceptions with how they feel others feel.

Another key finding is the relationship between satisfaction, quality of life and intention to recommend the event to others. Recommending the event to others is an important component of subjective well-being, as it can encourage people to attend together which can help people achieve a greater sense of belonging. Despite not being overly satisfied with the event as a whole or feeling that the event had contributed to their quality of life in any particular way, the respondents would, in general, recommend the event to others. Moreover, these variables show a high correlation, confirming previous findings that people who are more satisfied with the elements of an event are more likely to support and recommend it to others (Kaplanidou *et al.*, 2013; Li & Wan, 2017; Ouyang, Gursoy & Chen, 2019).

Recommendations and conclusion

This study contributes to the theoretical considerations of quality of life and to event management. A special contribution of the study is it analyses quality of life in the specific context of Christmas markets, a topic overlooked in previous studies. Conveying knowledge and making recommendations to the event's organisers and other stakeholders are especially important in terms of managerial implications. One recommendation to the event's organisers is to promote the authentic local culture through entertainment and cultural events, as this would help to enhance the competitiveness of both Advent in Zagreb and the entire destination in the tourism market. The results suggest that if residents feel that Advent in Zagreb encourages the preservation of the local culture, it is highly likely this would have a favourable effect on their quality of life. Another recommendation to the event's organisers is to seek out alternative solutions to traffic congestion, considering that free public transport only on weekends has proved ineffective. To boost their sales volume, product and service providers at the event are advised to either improve the quality of their products/services or reduce their prices, to make them more affordable to residents, as the respondents are partly satisfied with the offering but dissatisfied with high prices. A possible solution would be to cut prices, particularly when taking into consideration the average purchasing power of residents and tourists. The event organisers would also be advised to involve residents in the process of planning and holding the event, and in the promotion and integration of local cultural heritage, to achieve greater social cohesion and solidarity (Winchester & Rofe, 2005; Li & Wan, 2017) and ensure the best possible quality of life for the local community. It is generally accepted in previous studies (e.g. Đurkin & Perić, 2017; Đurkin & Wise, 2018; Gursoy *et al.*, 2019) that local authorities should consider residents' quality of life as a key factor when seeking to plan and organise a sustainable event, in particular an event that recurs annually.

The study focuses exclusively on residents' quality of life during the duration of the event Advent in Zagreb. To further analyse the effect of this event on residents' quality of life, it is recommended to conduct similar studies using a longitudinal approach, before and after the event. This would provide deeper insight into the current perception of residents' quality of life, regardless of the season, that is, regardless of the event (Han *et al.*, 2017; Bimonte, D'Agostino & Grilli, 2020; Ouyang, Gursoy & Chen, 2019). The test instrument (questionnaire) in future studies could include certain objective indicators of quality of life, such as respondents' income. Adding income to other socio-demographic variables, such as educational background or length of residence in the city, would help to better understand the level of quality of life achieved during the event (Moon *et al.*, 2019; Bimonte, D'Agostino & Grilli, 2020). In this regard, care should be taken to ensure that the sample is representative in all its characteristics, including a balanced gender structure which is not the case now. More complex methods such as regression or structural modelling would also provide additional insight into the complex relationship between variables. Finally, geographical determinants, as suggested by Han *et al.* (2017), could also be included in studies of residents' quality of life to explore whether there are any differences in residents' quality of life depending on the neighbourhood in which they live, and whether there are any differences in perceived quality of life among people living in parts of the city where event venues are located and those living in neighbourhoods with no event venues.

References

Bimonte, S., D'Agostino, A., & Grilli, G. (2020). Tourist season and determinants of residents' subjective well – being. *Annals of Tourism Research*. 81 (C), 102744.

Brida, J.G., Disegna, M., & Osti, L. (2013). Visitors' expenditure behaviour at cultural events: The case of Christmas markets. *Tourism Economics*. 19 (5), 1173–1196.

Brida, J.G., Meleddu, M., & Tokarchuk, O. (2017). Use value of cultural events: The case of the Christmas markets. *Tourism Management*. 59, 67–75.

Carneiro, M.J., & Eusebio, C. (2015). Host–tourist interaction and impact of tourism on residents' quality of life. *Tourism and Management Studies*. 11 (1), 25–34.

Colombo, A. (2016). How to evaluate cultural impacts of events? A model and methodology proposal. *Scandinavian Journal of Hospitality and Tourism*. 16 (4), 500–511.

Dagger, T.S., & Sweeney, J.C. (2006). The effect of service evaluations on behavioral intentions and quality of life. *Journal of Service Research*. 9 (1), 3–18.

Đurkin, J., & Perić, M. (2017). Organising for community-based tourism: Comparing attitudes of local residents and local tourism entrepreneurs in Ravna Gora, Croatia. *Local Economy, the Journal of the Local Economy Policy Unit*. 32 (7), 678–691.

Đurkin, J., & Wise, N. (2018). Managing community stakeholders in rural areas: Assessing the organisation of local sports events in Gorski kotar, Croatia. In: Clarke, A., & Jepson, A. (eds.) *Power, construction and meaning in festivals & events*. London: Routledge, pp. 185–200.

Gursoy, D., Ouyang, Z., Nunkoo, R., & Wei, W. (2019). Residents' impact perceptions of and attitudes towards tourism development: A meta-analysis. *Journal of Hospitality Marketing & Management*. 28 (3), 306–333.

Han, J., Wang, W., Zheng, C., & Zhang, J. (2017). Host perceptions of music festival impacts: Time and space matter? *Asia Pacific Journal of Tourism Research*. 22 (11), 1156–1168.

Iwasaki, Y. (2007). Leisure and quality of life in an international and multicultural context: What are major pathways linking leisure to quality of life? *Social Indicators Research*. 82, 233–264.

Kahraman, A., & Kazançoğlu, İ. (2019). Understanding consumers' purchase intentions toward natural-claimed products: A qualitative research in personal care products. *Business Strategy and the Environment*. 28 (6), 1218–1233.

Kaplanidou, K. (2012). The importance of legacy outcomes for Olympic games four summer host cities residents' quality of life: 1996–2008. *European Sport Management Quarterly*. 12 (4), 397–433.

Kaplanidou, K., Karadakis, K., Gibson, H., Thapa, B., Walker, W., Geldenhuys, S., & Coetzee, W. (2013). Quality of life, event impacts, and mega-event support among South African residents before and after the 2010 FIFA world cup. *Journal of Travel Research*. 52, 631.

Kersulić, A., Perić, M., & Wise, N. (2020). Assessing and considering the wider impacts of sport–tourism events: A research agenda review of sustainability and strategic planning elements. *Sustainability*. 12 (11), 4473.

Kim, K., Uysal, M., & Sirgy, M.J. (2012). How does tourism in a community impact the quality of life of community residents? *Tourism Management*. 36, 527–540.

Kim, W., Jun, H.M., Walker, M., & Drane, D. (2015). Evaluating the perceived social impacts of hosting large-scale sport tourism events: Scale development and validation. *Tourism Management*. 48, 21–32.

Li, X., & Wan, Y.K.P. (2017). Residents' support for festivals: Integration of emotional solidarity. *Journal of Sustainable Tourism*. 25 (4), 517–535.

Liang, Z.X., & Hui, T.K. (2016). Residents' quality of life and attitudes toward tourism development in China. *Tourism Management*. 57, 56–67.

Ma, S.C., & Kaplanidou, K. (2016). Legacy perceptions among host Tour de Taiwan residents: The mediating effect of quality of life. *Leisure Studies*. 36 (3), 423–437.

Marcher, A., Erschbamer, G., & Pechlaner, H. (2019). Escaping from the event? Residents' perception of Christmas markets in Northern Italy. *Event Management*. 23 (4–5), 599–611.

Moon, B.-Y., Yang, S.-H., Kim, K.-S., & Seo, W.-J. (2019). Resident-focused tourism policy: Perceived quality-of-life difference and its effect on expectations about and support for mega events. *Journal of Policy Research in Tourism, Leisure and Events*. 11 (2), 220–235.

Moscardo, G. (2009). Tourism and quality of life: Towards a more critical approach. *Tourism and Hospitality Research*. 9 (2), 159–170.

Musikanski, L., Rogers, P., Smith, S., Koldowski, J., & Iriarte, L. (2019). Planet happiness: A proposition to address overtourism and guide responsible tourism, happiness, well-being and sustainability in world heritage sites and beyond. *International Journal of Community Well-Being*. 2, 359–371.

Nofre, J., Giordano, I.E., Eldridge, A., Martins, J.C., & Sequera, J. (2018). Tourism, nightlife and planning: Challenges and opportunities for community liveability in La Barceloneta. *Tourism Geographies*. 20 (3), 377–396.

Ouyang, Z., Gursoy, D., & Chen, K.-C. (2019). It's all about life: Exploring the role of residents' quality of life perceptions on attitudes toward a recurring hallmark event over time. *Tourism Management*. 75, 99–111.

Pavluković, V., Armenski, T., & Alcántara-Pilar, M. (2017). Social impacts of music festivals: Does culture impact locals' attitude toward events in Serbia and Hungary? *Tourism Management*. 63, 42–53.

Perić, M. (2018). Estimating the perceived socio-economic impacts of hosting large-scale sport tourism Events. *Social Sciences*. 7 (10), 176.

Perles-Ribes, J.F., Ramón-Rodríguez, A.B., Moreno-Izquierdo, L., & Such-Devesa, M.J. (2020). Tourism competitiveness and the well-being of residents: A debate on registered and non-registered accommodation establishments. *European Journal of Tourism Research*. 24, 2406.

Richards, G. (2017). From place branding to placemaking: The role of events. *International Journal of Event and Festival Management*. 8 (1), 8–23.

Richards, G., de Brito, M.P., & Wilks, L. (eds.) (2013). *Exploring the social impacts of events*. London: Routledge.

Smith, A. (2012). *Events and urban regeneration*. London: Routledge.

Spirou, C. (2011). *Tourism and urban change: Cities in a global economy*. London: Routledge.

Uysal, M., Sirgy, M.J., Woo, E., & Kim, H.L. (2016). Quality of life (QOL) and well-being research in tourism. *Tourism Management*. 53, 244–261.

Winchester, H.P.M., & Rofe, M.W. (2005). Christmas in the 'valley of praise': Intersections of the rural idyll, heritage and community in Lobethal, South Australia. *Journal of Rural Studies*. 21 (3), 265–279.

Wise, N. (2016). Outlining triple bottom line contexts in urban tourism regeneration. *Cities*. 53, 30–34.

Wise, N., & Harris, J. (eds.) (2017). *Sport, events, tourism and regeneration*. London: Routledge.

Yolal, M., Gursoy, D., Uysal, M., Kim, H., & Karacaoğlu, S. (2016). Impacts of festivals and events on residents' well-being. *Annals of Tourism Research*. 61, 1–18.

Zhao, Y., & Wise, N. (2019). Evaluating the intersection between "green events" and sense of community at Liverpool's Lark Lane farmers market. *Journal of Community Psychology*. 47 (5), 1118–1130.

7 Can a "bean" spread well-being?

Francesc Fusté-Forné and Maria Pont-Llagostera

Introduction

Food is an identity marker that conveys the meanings attached to both cultural and natural environments. Food is constantly valorised through leisure and tourist activities and experiences (Kim & Eves, 2012; Everett, 2019; Brune *et al.*, 2020). Food as a social ingredient forms part of daily habits of people in the contemporary world, and food as a tourism attraction is growing its importance as both a travel motivation and a destination landmark (see Fusté-Forné, 2020). In this sense, culinary and gastronomy tourism refers to the discovery of local, cultural and natural idiosyncrasies through food (Long, 2004). Food is a source of well-being. In this chapter, a food event emerges as a significant practice where to gather physical and symbolic well-being through food, not only from a demand perspective. Specifically, this study analyses the impact of *Fira del Fesol* (Bean Fair) on the well-being of restaurant entrepreneurs in Santa Pau, a village that lies at the centre of the Garrotxa Volcanic Zone, home of a Natural Park in central Catalonia, north-eastern Spain.

Well-being is a crucial indicator to measure a country's prosperity (OECD, 2020) and the third UN Sustainable Development Goal (UNESCO, 2020) aims at "good health and well-being". Lee (2019, p. 2) affirms that "although researchers have no clear consensus on the definition of well-being, most agree that well-being encompasses several domains of human functioning, relating to issues of mental health, life satisfaction, and social functioning as well as practical aspects of quality of life". Well-being is a condition towards a "good life" (Diener, Scollon & Lucas, 2003; Ares *et al.*, 2016), which includes, among others, aspects such as satisfaction with work and social life (Ares *et al.*, 2016). Block *et al.* (2011) discussed the notion of food well-being and affirmed that people do not only eat to seek physical nourishment. In particular, Block *et al.* (2011, p. 6) defined food well-being as "a positive psychological, physical, emotional, and social relationship with food at both the individual and societal levels".

In this context, connections between humans and the environment also contribute to well-being (Hanna *et al.*, 2019). As Su and Zhang (2020) have underlined with regard to tea production and consumption, this research advocates that bean production and consumption are well-being driver factors. As teascapes do, beanscapes "can easily foster interactions between people, settings, and objects to enhance social well-being" (Su & Zhang, 2020, p. 5). Furthermore, planning and development of local and sustainable economic activities lead to "deeper human connections, well-being, and a greater sense of reciprocity with the natural environment" (Chassagne & Everingham, 2019, p. 1911). In the framework of "tastescapes of well-being", a bean provides a taste of local, which spread well-being.

Previous research has analysed the relationships between well-being and quality of life (Dolnicar, Yanamandram & Cliff, 2012) and also the specific role of food in improving the quality of life (Buiatti, 2011; Hernández-Mogollón, Di-Clemente & Campón-Cerro, 2020). Yolal *et al.* (2016) advocate that events contribute to the well-being of local residents. Following previous studies that have focused on the study of well-being from a community perspective, this chapter is also based on a local approach. Departing from the understanding of event tourism as an avenue towards sustainable development, this research adds empirical texture to the relationships between events and well-being through "food". Results reveal why restaurants are proudly offering bean-based recipes as part of a volcanic-based cuisine, and how, thanks to the fair, visitors can taste a portion of the territory through a "bean".

Literature review

This section is divided into two sub-sections. First, it discusses the contribution of food to well-being. Later, it approaches the concept of event tourism.

Shaping food significance on well-being

Previous research has investigated the influence of food on well-being and highlighted that food affects physical and phycological health, and contributes to social relationships (Ares *et al.*, 2016). Food acts as a "reference point" (Montanari & Staniscia, 2009) that promotes the cultural and natural identity of a place (Fusté-Forné, 2015; Back, Okumus & Tasci, 2020). Food is not only a basic need to sustain humans, but it also provides emotional and symbolic values, both individual and social (Fifita *et al.*, 2020). Food expresses identity (Long, 1998). To taste a food is to taste a culture, which does mean not only to explore an identity but also to remember unique and meaningful experiences (Fusté-Forné, 2020). In this sense, the creation of food experiences is the result of a collective effort, as it is exemplified in food festivals which often rely on local produce as a *leitmotiv* (Andersson, Mossberg & Therkelsen, 2017). This study is focused on a food event based on locally grown beans.

According to Lum and Ferrière le Vayer (2016, p. 4),

> foodways-related intangible heritage is a form of intergenerational communication or, put differently, the transmission over time of distinct food-related practices, representations, expressions, knowledge, and skills that members of a community, large or small, consider to be vital or significant to their cultural well-being or uniqueness.

As a consequence, foods and drinks have also a positive effect on residents' "sense of cultural belonging" (Andersson, Mossberg & Therkelsen, 2017). In this sense, Hjalager and Johansen (2013) reported that a growing demand for quality food services and added value food experiences is attached to the understanding of food as a crucial component of well-being. This can be specially applied to protected areas, where the environment represents an integral aspect for the configuration of foods and dishes.

The discovery of local foods as part of leisure and tourist practices is also based on social reasons (Tikkanen, 2007). In this context, slow food experiences must engage people with responsible and mindful relationships that foster not only resilience and sustainability but also social and ecological well-being (Fusté-Forné & Jamal, 2020). Traditional culinary heritages and practices, "which requires time and dedication, fits in the modern trend for 'slow food'. The time and care needed to prepare traditional dishes is a rarity in the modern world. At the same time, it so appears that the common table that symbolises harmony, unity and continuity of inheritance is so important that being separated from it can affect one's cultural identity and integrity" (Dominik, 2020, p. 123). Previous research has analysed local cuisine as a major factor towards regional development (Bessière, 1998; Kivela & Crotts, 2005), where the availability of a structured offer can lead to an increase in destination attractiveness (Lin, Pearson & Cai, 2011; Back, Okumus & Tasci, 2020). This includes, for example, food events (Getz, 1991; Hall & Sharples, 2003). In this context, Yolal *et al.* (2016) report that subjective well-being is "a product of individuals' emotional reactions to events, and their cognitive evaluations of satisfaction and fulfillment these events can bring to their life" (p. 3).

Furthermore, the development of slow food practices allows both residents and visitors to be active travellers and to participate in a co-created experience with local producers, where they build a dialogue that is a source of cultural exchange. Thus, they can understand where food comes from, how it is grown, distributed and consumed, and how it affects personal and ecological health (Fusté-Forné & Jamal, 2020). Here, consumption of organic foods contributes to long-term physical, psychological and social well-being (Lee, 2019). As discussed later in the chapter, well-being is based not only on functional properties of food but also on its embedded symbolism (Fifita *et al.*, 2020) – a representation of sense of place and *terroir* (Trubek, 2008).

Understanding food in event tourism

Getz has largely reviewed the significance of events, and he has built on the notion of event tourism (see Getz, 1991). Getz and Page (2016) affirm that event tourism has experienced an impressive growth since 2008. In this sense, events are both a motivator factor for tourists and a significant element in the planning and development of destination management and marketing. Among the benefits of event tourism, these authors state that it is a form of tourism that contributes to face the seasonality of tourism demand and to expand the spatial influence of tourism in a territory.

Events make places more liveable and generate a permanent, shared and inherited legacy. Here, festivals and other cultural celebrations "are links to the preservation of traditions, group identity building and legitimation, communitas among subcultural groups, generation of social and cultural capital, and the value of celebrations in establishing place identity and civic pride" (Getz & Page, 2016, p. 606). Foods are particularly attractive for event planning and development, and food events serve as tools for the communication of both tangible and intangible heritages. Previous research highlights the significance of food events towards the promotion of regional identity (see, e.g. Hashimoto & Telfer, 2008). In this sense, a food event exemplifies a process which contributes to the construction of social networks (Putnam, 2000) that may facilitate cultural and social well-being.

Events are a salient attribute for individuals and societies, and they succeed even in periods of crisis because events satisfy basic human needs (Getz, 2012, cited by Seraphin, 2020). This is also worthy of attention in peripheries, where food-based festivals and events are crucial for the economic sustainability of rural communities (Getz & Robinson, 2014). While a wide range of wine events have been celebrated worldwide in recent decades (see Yeoman *et al.*, 2012) as a cornerstone to develop event tourism, this has also extended to other types of food events based on examples of specific segments of food tourism such as asparagus (Lewis, 1997), cheese (Folgado-Fernández, Di-Clemente & Hernández-Mogollón, 2019) or mushrooms (Fusté-Forné, 2019), among others.

Food events are examples of food-based leisure and tourism activities which protect and promote both artisanal agriculture and sustainable communities (Star, Rolfe & Brown, 2020). Food festivals are growingly used to brand destinations (see Lee & Arcodia, 2011; Yang *et al.*, 2020). As a consequence, advertising and marketing strategies should highlight the significance of organic food consumption as an avenue towards physical, psychological and social well-being (Lee, 2019). Also, Yang *et al.* (2020) report that the development of an event implicitly includes the promotion of a destination and therefore *unique* knowledge about local communities is gathered: La Mancha means cheese, Valencia means paella and Santa Pau means beans. Santa Pau *Bean Fair* let residents and visitors navigate into the identity of the place, taste a volcano gastronomy and engage with tangible and intangible heritages of local people and local landscapes.

Study method

Bean Fair in Santa Pau is celebrated every year (in 2020, its 30th edition was organised) in January, and it implies the dedicated participation of local restaurants that deliver a range of small dishes (*tapas*) to visitors from individual stands located in the town centre (Figures 7.1 and 7.2). Recipes must use Santa Pau beans, a product that counts on a Protected Designation of Origin (PDO). This label acknowledges a quality product, the land where it grows, and people who take care of it. Beans from Santa Pau are closely linked to the land and they uniquely showcase the volcanic *terroir* of this area, La Garrotxa. The region is part of a Natural Park which represents one of the largest volcanic landscapes in

Figure 7.1 Restaurant stands at *Bean Fair*, Santa Pau

Source: Maria Pont-Llagostera.

Europe (La Garrotxa Volcanic Zone Natural Park, 2012). Drawing on nine semi-structured interviews with eight of the restaurant owners that participate in the fair, and its organiser, this research employs a case study approach (Yin, 2009) to discuss how and why their engagement with beans and *Bean Fair* has an impact on their well-being.

Profile of restaurants, observed in Table 7.1, reveals that all the interviewees have been involved with the territory for a long period, with the oldest restaurant being established in 1931. These restaurants are small, familiar and rural businesses which protect and promote a local-based cuisine. Some of them have

Figure 7.2 Decoration of the restaurant stands at *Bean Fair*, Santa Pau

Source: Maria Pont-Llagostera.

Table 7.1 Profile of restaurants

Restaurant	*Genre of the Interviewee*	*Age*	*Restaurant Opening (Year)*	*Participation in* Bean Fair *Since (Year)*
1	Male	54	1931	1991
2	Male	36	1991	1992
3	Female	82	1985	1992
4	Female	59	1992	1994
5	Male	57	1983	1992
6	Male	40	2000	2003
7	Male	48	1990	2005
8	Male	58	1993	1997

participated in *Bean Fair* since it began in 1991. All the interviews were conducted during the month of August 2020, and they lasted between 15 and 30 minutes. Interviews started with basic demographic questions to later focus on the meanings that participants attach to a "bean" as a symbol of the territory. Also, interviews aimed to explore the presence of beans in restaurant menus and the reasons to participate in *Bean Fair*. In particular, interviewees were questioned about the impact of the event on their individual and social well-being. The interviews were conducted in Catalan language. They were transcribed and analysed, in their original language, using a qualitative data analysis software for category development. Results are described in the next section.

Results and discussion

Bean Fair in Santa Pau (Catalonia, Spain) has been held since 1991. Its origin begins with the interest built by local bean growers towards the recovery and protection of the Santa Pau bean as a local ingredient. Its culinary uses emerged as a platform to secure the reappreciation of beans, which rapidly resulted in the involvement of restaurants with the organisation of a "fesolada" (big cooking of beans) which is the antecedent of the current festival. According to the event organiser, "in Catalonia there was only the honey fair, and 'calçotada' [big cooking of scallions] in Valls. Santa Pau was one of the pioneering municipalities in including gastronomy in a fair". This reveals the critical role of food in local lifestyles, and how residents are engaged with the promotion of this legume.

Bean Fair is co-organised by Santa Pau Council, Santa Pau Bean Growers Association and Sant Antoni Fair Commission, with the collaboration of *Cuina Volcànica* (Volcanic Cuisine), which is a group of restaurants rooted in the Garrotxa region. They offer a passionate cuisine based on local products, such as corn, buckwheat, black turnips, black potato, and products such as Santa Pau beans, pork and scarlets or truffles which make the landscape be reflected in dishes (Cuina Volcànica, 2020). In this sense, practices in rural areas are based on a direct contact between people and nature (see Lane, 1994). According to Boukas (2019), the utilisation and promotion of rural resources protect rural traditions and promote a sustainable development which "can also affect the well-being of their local residents" (p. 64). This is even more relevant when the natural setting is part of a protected area (Hjalager & Johansen, 2013).

The value of the territory

The recovery and promotion of beans have resulted in the popularity of this small town in Central Catalonia. All the restaurants have acknowledged Santa Pau beans as the most representative product of Santa Pau identity, its "star" product, an "emblem" of the land [R6]. Beans show a clear connection between nature and culture since:

> beans are grown in a very special water and soil, which contains a lot of minerals, and they have a special taste and shape. We would never have imagined

that having such a simple product will be very appreciated and greatly revalorised. This product has given its name to the volcanic cuisine.

[R1]

Another interviewee underlined that:

> it's an original and a very special product for us, very good and healthy, with very different characteristics from another beans, it's unique. I've always thought that if a product from the land has quality, it is culture. That's the reason why we have to keep it and preserve, strive for it and love it. It comes from nature.
>
> [R3]

This demonstrates that a close relationship between humans and nature is a primary source of well-being, where "place" becomes a critical attraction factor. Silver and Grek-Martin (2015) understand the sense of place as a mix of emotions and experiences which are transmitted in relation to a specific locality. Furthermore, Smith (2015) defines it as the amalgam of landscapes, heritage and social structures which contribute towards the comprehension of a local identity (Antrop, 2005; Haven-Tang & Jones, 2005; Berno, Laurin & Maltezakis, 2014; Fusté-Forné, 2020). In this context, the notion of sense of place "encapsulates the idea that a particular interplay of geography, history and human factors imbues foods with a particular taste that cannot be recreated elsewhere" (Musgrave, 2009, p. 158).

All the participants highlighted the healthy values of Santa Pau beans, and its contribution not only to economic development but also to ecological and social advancement. In this sense, restaurants indicated the cultivation of beans is an artisanal and manual process which acknowledges a genuine product, a unique land, and local people who take care of the territory and its traditions. This understanding results in a positive effect of bean production for locals (i.e. identity) and a huge promotion of Santa Pau brand to visitors and tourists:

> beans represent the village of Santa Pau, thanks to the involvement of the peasantry. In addition, a product has named after a village as a representative of the gastronomy of La Garrotxa and its volcanic cuisine, and together we have achieved that this product is now in the market and is popular all over the country.
>
> [R5]

Santa Pau bean portrays Santa Pau, and the volcanic environment and soil where it lies, and "it helps to consolidate an area, a town and a territory" [R8]. In this sense, the attraction of volcanic areas relies on the geological heritage of their unique landscape features (Dowling & Newsome, 2006). Economic resources in volcanic areas mostly come from tourism (Sigurdsson & Lopes-Gautier, 2000), where food is regarded as a significant heritage and a meaningful attraction. While the crucial importance that the configuration of the land and the landscape – the geology – have for foods and the taste of foods – food is a conveyor of the sense of place, previous research alerts that a lot of work is still required to further engage residents and visitors in co-creating food-based experiences in natural environments (Hjalager & Johansen, 2013).

Santa Pau beans in restaurant menus

It is obvious that Santa Pau beans are largely included in restaurant menus. Moreover, tastings and demonstrations are carried out periodically. Some of the restaurants offer beans as part of their iconic dishes, such as the "Santa Pau beans with sausage and bacon steaks" [R2]. Another restaurant exposed a sense of seasonality in bean pairing with other foods which are traditional in summer or winter periods. For example, beans with mushrooms [R8] which are traditionally picked during autumn. In this sense, seasonal dishes do increase not only the regional economic impact of local produce but also the effects of the relationship between food and well-being, both physically and symbolically – food in the right place, in the right moment. All the restaurants highlighted that beans are available in the menus on a daily basis and even one of them stated that its menu offers up to 8–10 bean-based dishes [R5].

Thus, local produce holds a powerful capacity to appreciate local lifestyles and traditions in a framework of slow co-production and co-consumption. According to Fusté-Forné and Jamal (2020), slow food supports degrowth and, through regional-based producers, facilitates sustainable livelihoods and well-being – individual and communal. Slow food practices help destinations to work towards the consecution of the UNESCO's Sustainable Development Goals (UNESCO, 2020).

Participant number 3 exemplifies the integration of beans in restaurant menus. First, during the winter season, "we offer mixed beans of Santa Pau with sausage of 'perol', which is the most special dish and the first recipe we prepared to participate in the fair". Second, during the summer season, "we use to create a plate with beans, fresh vegetables from our organic garden and cod fish". They also prepare a bean-based dessert, which is the most representative dish of the restaurant participation at the event (Figures 7.3 and 7.4):

> we even prepare desserts with beans, the most special one is 'volcanets' [*small volcanoes*] muffins of beans. This is our star dish. It's an exclusive desert, which you can only taste during the fair, not at the restaurant.

In the process of attributing tourism value to local food, especially in rural areas, we must consider the adoption of slow food enhances an equitable utilisation of resources that secures a long-term viability of rural environments (Boukas, 2019). This happens with activities such as visiting farmers' markets, whose role does not only relies on selling, but it is also as a facilitator of community building, an avenue to support cultural values, and a strategy to promote well-being (Belen & Ripoll, 2020).

Figure 7.3 Elaboration of "volcanets" at the restaurant, Santa Pau

Source: Maria Pont-Llagostera.

Thus, some restaurants design special dishes to display at the fair, others repeat the culinary elaborations every year, but all of them showcase the significance of this cultural and social event as a source of memorability and uniqueness.

Bean Fair: a must-do, a must-visit

In a similar vein to what Su and Zhang explored with regard to teascapes as an example of tastecape, these results confirm that "well-being is embodied in doings and enactments. This emphasis on well-being via tastescape is part

Figure 7.4 Delivery of "volcanets" at *Bean Fair*, Santa Pau

Source: Maria Pont-Llagostera.

of a wide shift made through tourism which allows tourists and hosts to maintain well-being elsewhere and elsewhen" (Su & Zhang, 2020, p. 18). In particular, food contributes to both individual and social well-being (Batat *et al.*, 2019), especially when it is communicated through small and sustainable tourism projects (see, e.g. Hatipoglu, Ertuna & Salman, 2020): a food fair. In this sense, collaboration and cooperation between food producers and restaurants in peripheral areas is a crucial issue to the success of events such as the *Bean Fair* and its consolidation not only as a cultural event but also as a tourist event. As observed in Table 7.1, the vast majority of restaurants have been involved with the fair since its first editions. Why?

> First of all, for the love to the village and because we have always had trust on beans. I also think that the product must be preserved, valued and respected. Get to know each other, make cohesion and cooperate with other participants it's a fantastic philosophy.
>
> [R3]

Restaurants expressed they are always ready to support local-based events organised by local people which helps the territory in a broad sense. They are also local actors who are attached to the place. One of the interviewees affirmed that "we appreciate a lot Santa Pau and its people, we are part of them. When they ask us to participate at the fair we always say yes, we are always open" [R4].

At the beginning, "the fair was promoted to reduce the seasonality during the winter season" [R1], affirmed one of the interviewees. After its 30th edition celebrated in 2020, most of the restaurants who participate in the event acknowledged the co-created bonds between local stakeholders, event organisers and tourists. For example, "it's nice to see how so many visitors come and come year after year and they remember our 'volcanets'" [R3]. The fair supports the creation of bridges: producers deliver beans to restaurants, and each restaurant cooks them in its own way. After that, both residents and visitors enjoy a wide range of dishes made of beans. Finally, many respondents also reported the role of *Bean Fair* as "a platform to publicise the product and sell" [R8]. They achieve a greater visibility because of the fair and, ultimately, it is a strategy to increase restaurant revenue.

From a "simple" participation in a fair to a "complete" promotion of a territory

Bean Fair distinctively contributes to the promotion of local cuisine. This is perceived and valued by local restaurants, as mentioned earlier, and they understand the fair as an avenue towards social cohesion and common exhibition of the land – via Santa Pau beans. One interviewee reported that:

> *Bean Fair* is a successful event where participants like to attend even if it's very cold, bad weather or economic crisis. Many people associate Santa Pau with these words: beans, green, nature and volcanoes.
>
> [R1]

Bean Fair is "a speaker for local cuisine, and its positive for people and village" [R2]. While the event helps both restaurants and tourism industry in the area, it is primarily a local co-shared celebration [R7]. This perception reveals the identification of beans as a festive food. Moreover, beans are available all over the year, which means they can be always included in restaurant menus and in family meals.

Due to the diverse offer of dishes – some of them being different than regular restaurants' offer, the fair "helps to spread the word about the cuisine we have. We have people from here and abroad, and the product is promoted and recognised more and more" [R5], as discussed earlier. In addition, some of the interviewees warned that food fairs have quickly spread over the territory:

> earlier, fairs were very seasonally marked, but nowadays there are plenty of events and everything can be turned into a fair which may result in a devaluation of some products. There are fairs which aren't authentic, which are not rooted in the territory, but the Santa Pau *Bean Fair* does, and this is its virtue.
>
> [R8]

The fair engages residents and tourists with a local food and, according to the proud participants, contributes to tourist intention to visit, recommend and revisit. Why? Three reasons are outlined: the pure landscape, the passionate people and every year a *unique* recipe are factor drivers to the bean experience and its well-being share. In this sense, relationships between foods and events serve to create a narrative of local produce and local people. According to Palupi and Abdillah (2019, p. 309) "culture is told to the community through storytelling to produce history. This story contains information about food: why it should be made, how it is presented, raw materials needed, and standard rules needed that should not be violated".

As a consequence, food-based events have implications for both identity and ways of life (Lau & Li, 2019; Yang *et al.*, 2020), and they engage various generations in the process of learning the meanings attached to culinary heritages. This is especially relevant for younger generations who, in line with the slow food movement, "can discover alternatives to the globalised gastronomy and experience the potential of tradition as the store of cultural memory" (Dominik, 2020, p. 123). Later, this knowledge is shared with visitors and tourists who can celebrate *authentic* products that rely on *unique* landscapes (Hjalager & Johansen, 2013; Berno & Fusté-Forné, 2019).

Conclusion

This chapter has explored the motivations of local stakeholders to build a local-based culinary offer as part of food and event tourism in volcanic environments, and it demonstrates how a bean can promote well-being among residents. A central identity role of beans is revealed. All the participants highlighted the uniqueness of Santa Pau bean and the fair as a transmission chain of culinary heritages – focused on distinctive cultural and natural idiosyncrasies. Also, cooperation among restaurants and collaboration between food producers and other stakeholders are examples of how food contributes to the sustainable development of rural areas. Moreover, this is extended to other regional producers, as the organiser of the event exemplified: "every year a different guest product comes from other parts in Catalonia, for example Viladrau chestnut, and then culinary exchanges are made with other places like Valls and its 'calçotada'". Drawing on the attraction of local foods and gastronomic traditions in Catalonia, these bridges exemplify a network of food producers who pursue a greater economic and social well-being.

Furthermore, events represent well-established communities of practice (Fusté-Forné & Nguyen, 2018). This is the case of *Bean Fair* in Santa Pau where the relationships between food and tourism secure economic, social and environmental sustainability and strength the region's identity (Everett & Aitchison, 2008). This study has particularly built on the linkages between local events and individual and community well-being in protected areas through the case of "beans".

Theoretical and practical implications of this research are significant for event tourism management, and marketing in natural and rural areas that feature volcanoes as meaningful landmarks and contribute to the life satisfaction of people who live and work in the region. Especially, a perfect mixture is retained when events are focused on local produce. In this sense, "events have become a large focus for economic development in rural and regional areas. Tourism has been a key driver of food festivals, where they play a role in attracting visitors and generating expenditure in the local economy" (Star, Rolfe & Brown, 2020, p. 325). As discussed earlier, economic development is only a single factor. Further research also needs to explore the crucial role of beans towards the well-being of visitors and tourists (Baccarani & Cassia, 2017) and scrutinise differences among individual cultural backgrounds (Ares *et al.*, 2016). While the huge effects of the coronavirus outbreak on service industries (see Gössling, Scott & Hall, 2020; Williams, 2020), academics and practitioners are urged to further analyse the relationship between food, well-being and events in post-COVID tourism since events will remain at the core centre of human, nature and tourist experiences.

References

Andersson, T.D., Mossberg, L., & Therkelsen, A. (2017). Food and tourism synergies: Perspectives on consumption, production and destination development. *Scandinavian Journal of Hospitality and Tourism*. 17 (1), 1–8.

Antrop, M. (2005). Why landscapes of the past are important for the future. *Landscape and Urban Planning*. 70 (1–2), 21–34.

Ares, G., Giménez, A., Vidal, L., Zhou, Y., Krystallis, A., Tsalis, G., Symoneaux, R., Cunha, L.M., Pinto de Moura, A., Claret, A., Guerrero, L., Cardello, A.V., Wright, A., Jefferies, L., Lloyd, M., Oliveira, D., & Deliza, R. (2016). Do we all perceive food-related wellbeing in the same way? Results from an exploratory cross-cultural study. *Food Quality and Preference*. 52, 62–73.

Baccarani, C., & Cassia, F. (2017). Evaluating the outcomes of service ecosystems: The interplay between ecosystem well-being and customer well-being. *The TQM Journal*. 29 (6), 834–846.

Back, R.M., Okumus, B., & Tasci, A.D. (2020). Culinary fans vs culinary critics: Characteristics and behavior. *International Hospitality Review*. 34 (1), 41–60.

Batat, W., Peter, P.C., Moscato, E.M., Castro, I.A., Chan, S., Chugani, S., & Muldrow, A. (2019). The experiential pleasure of food: A savoring journey to food well-being. *Journal of Business Research*. 100, 392–399.

Belen Yanotti, M., & Ripoll Gonzalez, L. (2020). *Food, community and wellbeing: An exploration of Harvest Launceston farmers' market*. Launceston: University of Tasmania.

Berno, T., & Fusté-Forné, F. (2019). Imaginaries of cheese: Revisiting narratives of local produce in the contemporary world. *Annals of Leisure Research*, 1–19.

Berno, T., Laurin, U., & Maltezakis, G. (2014). The special role of agriculture in food tourism. In: Wolf, E. (ed.) *Have fork will travel: Handbook for food tourism*. Seattle, WA: World Food Travel Association, pp. 105–114.

Bessière, J. (1998). Local development and heritage: Traditional food and cuisine as tourist attractions in rural areas. *Sociologia Ruralis*. 38 (1), 21–34.

Block, L.G., Grier, S.A., Childers, T.L., Davis, B., Ebert, J.E., Kumanyika, S., Laczniak, R.N., Machin, J.E., Motley, C.M., Peracchio, L., Pettigrew, S., Scott, M., & Van Ginkel Bieshaar, M.N.G. (2011). From nutrients to nurturance: A conceptual introduction to food well-being. *Journal of Public Policy and Marketing*. 30 (1), 5–13.

Boukas, N. (2019). Rural tourism and residents' well-being in Cyprus: Towards a conceptualised framework of the appreciation of rural tourism for islands' sustainable development and competitiveness. *International Journal of Tourism Anthropology*. 7 (1), 60–86.

Brune, S., Knollenberg, W., Stevenson, K.T., Barbieri, C., & Schroeder-Moreno, M. (2020). The influence of agritourism experiences on consumer behavior toward local food. *Journal of Travel Research*, 1–15.

Buiatti, S. (2011). Food and tourism: The role of the "slow food" association. In: Sidali, K., Spiller, A., & Schulze, B. (eds.) *Food, agriculture and tourism*. Berlin: Springer, pp. 92–101.

Chassagne, N., & Everingham, P. (2019). Buen Vivir: Degrowing extractivism and growing wellbeing through tourism. *Journal of Sustainable Tourism*. 27 (12), 1909–1925.

Cuina Volcànica. (2020). *Del paisatge a la taula*. Available from www.cuinavolcanica.cat [Accessed 10 August 2020].

Diener, E., Scollon, C.N., & Lucas, R.E. (2003). The evolving concept of subjective well-being: The multifaceted nature of happiness. *Advances in Cell Aging and Gerontology*. 15, 187–219.

Dolnicar, S., Yanamandram, V., & Cliff, K. (2012). The contribution of vacations to quality of life. *Annals of Tourism Research*. 39 (1), 59–83.

Dominik, P. (2020). Food tourism as a tool for intergenerational transfer of culinary traditions. *Geography and Tourism*. 8 (1), 115–124.

Dowling, R.K., & Newsome, D. (2006). *Geotourism*. London: Routledge.

Everett, S. (2019). Theoretical turns through tourism taste-scapes: The evolution of food tourism research. *Research in Hospitality Management*. 9 (1), 3–12.

Everett, S., & Aitchison, C. (2008). The role of food tourism in sustaining regional identity: A case study of Cornwall, South West England. *Journal of Sustainable Tourism*. 16 (2), 150–167.

Fifita, I.M., Seo, Y., Ko, E., Conroy, D., & Hong, D. (2020). Fashioning organics: Wellbeing, sustainability, and status consumption practices. *Journal of Business Research*. 117, 664–671.

Folgado-Fernández, J.A., Di-Clemente, E., & Hernández-Mogollón, J.M. (2019). Food festivals and the development of sustainable destinations. The case of the cheese fair in Trujillo (Spain). *Sustainability*. 11 (10), 2922.

Fusté-Forné, F. (2015). Cheese tourism in a world heritage site: Vall de Boí (Catalan Pyrenees). *European Journal of Tourism Research.* 11, 87–101.

Fusté-Forné, F. (2019). Seasonality in food tourism: Wild foods in peripheral areas. *Tourism Geographies*, 1–21.

Fusté-Forné, F. (2020). Savouring place: Cheese as a food tourism destination landmark. *Journal of Place Management and Development.* 13 (2), 177–194.

Fusté-Forné, F., & Jamal, T. (2020). Slow food tourism: An ethical microtrend for the Anthropocene. *Journal of Tourism Futures*, 1–6.

Fusté-Forné, F., & Nguyen, T. (2018). Communities of practice, identity and tourism: Evidence on cultural heritage preservation in world heritage sites. *Almatourism-Journal of Tourism, Culture and Territorial Development.* 9 (18), 1–22.

Getz, D. (1991). *Festivals, special events, and tourism.* New York: Van Nostrand Reinhold.

Getz, D. (2012). *Event studies. Theory, research and policy for planned events.* London: Routledge.

Getz, D., & Page, S.J. (2016). Progress and prospects for event tourism research. *Tourism Management.* 52, 593–631.

Getz, D., & Robinson, R.N. (2014). Foodies and food events. *Scandinavian Journal of Hospitality and Tourism.* 14 (3), 315–330.

Gössling, S., Scott, D., & Hall, C.M. (2020). Pandemics, tourism and global change: A rapid assessment of COVID-19. *Journal of Sustainable Tourism*, 1–20.

Hall, C.M., & Sharples, L. (2003). The consumption of experiences or the experience of consumption? An introduction to the tourism of taste. In: Hall, C.M., Sharples, L., Mitchell, R., Macionis, N., & Cambourne, B. (eds.) *Food tourism around the world. Development, management and markets.* Oxford: Butterworth Heinemann, pp. 13–36.

Hanna, P., Wijesinghe, S., Paliatsos, I., Walker, C., Adams, M., & Kimbu, A. (2019). Active engagement with nature: Outdoor adventure tourism, sustainability and wellbeing. *Journal of Sustainable Tourism.* 27 (9), 1355–1373.

Hashimoto, A., & Telfer, D.J. (2008). From sake to sea urchin: Food and drink festivals and regional identity in Japan. In: Hall, C.M., Sharples, L., Mitchell, R., Macionis, N., & Cambourne, B. (eds.) *Food tourism around the world. Development, management and markets.* Oxford: Butterworth Heinemann, pp. 249–278.

Hatipoglu, B., Ertuna, B., & Salman, D. (2020). Small-sized tourism projects in rural areas: The compounding effects on societal wellbeing. *Journal of Sustainable Tourism*, 1–23.

Haven-Tang, C., & Jones, E. (2005). Using local food and drink to differentiate tourism destinations through a sense of place: A story from Wales-dining at Monmouthshire's great table. *Journal of Culinary Science and Technology.* 4 (4), 69–86.

Hernández-Mogollón, J.M., Di-Clemente, E., & Campón-Cerro, A.M. (2020). Culinary travel experiences, quality of life and loyalty. *Spanish Journal of Marketing-ESIC*, 1–22.

Hjalager, A.M., & Johansen, P.H. (2013). Food tourism in protected areas – sustainability for producers, the environment and tourism? *Journal of Sustainable Tourism.* 21 (3), 417–433.

Kim, Y.G., & Eves, A. (2012). Construction and validation of a scale to measure tourist motivation to consume local food. *Tourism Management.* 33 (6), 1458–1467.

Kivela, J., & Crotts, J.C. (2005). Gastronomy tourism: A meaningful travel market segment. *Journal of Culinary Science and Technology.* 4 (2–3), 39–55.

La Garrotxa Volcanic Zone Natural Park. (2012). *Volcanoes. A field guide to La Garrotxa Volcanic Zone.* Manresa, Spain: Parc Natural de la Zona Volcànica de la Garrotxa.

Lane, B. (1994). What is rural tourism? *Journal of Sustainable Tourism.* 2 (1–2), 7–21.

Lau, C., & Li, Y. (2019). Analyzing the effects of an urban food festival: A place theory approach. *Annals of Tourism Research*. 74, 43–55.

Lee, H.J. (2019). Does consumption of organic foods contribute to Korean consumers' subjective well-being? *Sustainability*. 11 (19), 5496.

Lee, I., & Arcodia, C. (2011). The role of regional food festivals for destination branding. *International Journal of Tourism Research*. 13 (4), 355–367.

Lewis, G.H. (1997). Celebrating asparagus: Community and the rationally constructed food festival. *The Journal of American Culture*. 20 (4), 73–78.

Lin, Y.C., Pearson, T.E., & Cai, L.A. (2011). Food as a form of destination identity: A tourism destination brand perspective. *Tourism and Hospitality Research*. 11 (1), 30–48.

Long, L.M. (1998). Culinary tourism: A folkloristic perspective on eating and otherness. *Southern Folklore*. 55 (3), 181–204.

Long, L.M. (2004). *Culinary tourism*. Lexington, KY: University Press of Kentucky.

Lum, C.M.K., & Le Vayer, M.D.F. (2016). *Urban foodways and communication: Ethnographic studies in intangible cultural food heritages around the world*. Lanham, MD: Rowman and Littlefield.

Montanari, A., & Staniscia, B. (2009). Culinary tourism as a tool for regional re-equilibrium. *European Planning Studies*. 17 (10), 1463–1483.

Musgrave, S. (2009). Grain elevated: The fall and rise of red fife wheat. In: Cooke, N. (ed.) *What's to eat? Entrees in Canadian food history*. Montreal: McGill-Queen's University Press, pp. 145–164.

OECD. (2020). *Measuring well-being and progress: Well-being research*. Available from www.oecd.org/statistics/measuring-well-being-and-progress.htm [Accessed 1 August 2020].

Palupi, S., & Abdillah, F. (2019). Local cuisine as a tourism signature. *Bridging Tourism Theory and Practice*. 11, 299–312.

Putnam, R.D. (2000). *Bowling alone. The collapse and revival of American community*. New York: Simon and Schuster.

Seraphin, H. (2020). COVID-19: An opportunity to review existing grounded theories in event studies. *Journal of Convention and Event Tourism*, 1–33.

Sigurdsson, H., & Lopes-Gautier, R. (2000). Volcanoes and tourism. In: *Encyclopedia of Volcanoes*. San Diego, CA: Academic Press, pp. 1283–1299.

Silver, A., & Grek-Martin, J. (2015). "Now we understand what community really means": Reconceptualizing the role of sense of place in the disaster recovery process. *Journal of Environmental Psychology*. 42, 32–41.

Smith, S. (2015). A sense of place: Place, culture and tourism. *Tourism Recreation Research*. 40 (2), 220–233.

Star, M., Rolfe, J., & Brown, J. (2020). From farm to fork: Is food tourism a sustainable form of economic development? *Economic Analysis and Policy*. 66, 325–334.

Su, X., & Zhang, H. (2020). Tea drinking and the tastescapes of wellbeing in tourism. *Tourism Geographies*, 1–21.

Tikkanen, I. (2007). Maslow's hierarchy and food tourism in Finland: Five cases. *British Food Journal*. 109 (9), 721–734.

Trubek, A.B. (2008). *The taste of place: A cultural journey into terroir*. Berkeley, CA: University of California Press.

UNESCO. (2020). *UNESCO and sustainable development goals*. Available from https://en.unesco.org/sustainabledevelopmentgoals [Accessed 1 August 2020].

Williams, C.C. (2020). Impacts of the coronavirus pandemic on Europe's tourism industry: Addressing tourism enterprises and workers in the undeclared economy. *International Journal of Tourism Research*, 1–10.

Yang, F.X., Wong, I.A., Tan, X.S., & Wu, D.C.W. (2020). The role of food festivals in branding culinary destinations. *Tourism Management Perspectives*. 34, 1–11.

Yeoman, I., Robertson, M., Ali-Knight, J., Drummond, S., & McMahon-Beattie, U. (2012). *Festival and events management*. London: Routledge.

Yin, R.K. (2009). *Case study research: Design and methods*. Thousand Oaks, CA: Sage.

Yolal, M., Gursoy, D., Uysal, M., Kim, H.L., & Karacaoğlu, S. (2016). Impacts of festivals and events on residents' well-being. *Annals of Tourism Research*. 61, 1–18.

8 'Whose way? my way!' evaluating well-being in the co-creation of 'end of life' events

Amrit Singh, Amanda Payne and Paul Fallon

Introduction

This chapter considers end-of-life events (also referred to as funerals) through the 'gaze' of more general event management concepts and practices. The authors support Laws and Deverell's (2019, p. 328) proposition that end-of-life events are notable by their 'almost complete absence' in the academic literature to date and, therefore, propose the ideas presented here as a springboard for further detailed consideration by event practitioners, teachers, students and researchers.

Funerals represent an interesting dichotomy for the events industry and its study, explaining both their marginalisation and signifying their potential. On the one hand, they are the one private event in which we will all share – death is the 'common denominator' for all of us. On the other hand, death usually signifies a difficult and stressful period (Bisconti, Bergman & Boker, 2004), and so funerals do not share the 'positive celebratory character' (Laws & Deverell, 2019), of other significant personal events such as weddings and birthdays. This chapter also argues that professionals working in these contexts – in this case, focusing on funeral arrangers (who arguably have the 'highest profile') to venue managers (who arguably go 'under the radar' in terms of their potentially significant role) also represent a dichotomy. Like the emergency services, they are essential when we need them but given little consideration when we don't (which is most of the time). Such marginalisation is also reflected in the literature to date.

Therefore, the purpose of this chapter is to emphasise the unique role which end-of-life occasions and those that organise them play in the general well-being of society and the individuals in it and also to start to understand the roles – and related well-being – of those working in this industry on behalf of the rest of us. There are three main objectives. Firstly, to give a 'voice' to event managers working within this context and thereby shed light on their labours and experiences. This represents a perspective seldom captured, even in the media during the challenging times of COVID-19. The second objective is to consider the applicability of wider event management concepts to end-of-life events and to propose a specific contribution to event management theory and practice via the CRADLES model which emphasises the importance and diversity of 'soft' service skills in this context. Finally, the authors wish to emphasise the significance of, and opportunities offered by, end-of-life occasions for the events industry and its workforce, and hopefully provide some guidance for skills development in this context.

The inspiration for the chapter comes from the main author, who started studying event management at undergraduate degree level to become a wedding planner and who, by the time of graduation, had embarked on a career as a funeral arranger. On the face of it, this may seem quite a transformation – although as the discussion indicates, given the nature of work and skills required, it does not in fact represent a major 'ship-jumping' exercise.

Due to the nature of the secondary data (including related academic literature) and the primary data sources and contexts included here, the discussion is oriented to the UK and Christian perspectives on funerals. However, it must be noted that this is not intentionally restrictive and exclusive – indeed the writing team includes those brought up in both non-Christian and Christian systems. Therefore, the reader is also encouraged to consider whether the ideas presented here are more widely generalisable.

Literature review

End-of-life events such as funerals, and therefore the various managers and service providers associated with them, are often 'overlooked'. Similar to our relationships with the emergency services, we tend not to think about them until we need them and then our need is both major and urgent. It is usually, but not exclusively, also short-lived. Notwithstanding this marginalisation, end-of-life events make a unique and significant contribution to the well-being of both society overall and the individuals within it (O'Rourke, Spitzberg & Hannawa, 2011). The fact that their *raison d'etre* equates to death is fundamental to the discussion here. Death is difficult for us to deal with – it is highly stressful (Bisconti, Bergman & Boker, 2004; Sutin *et al.*, 2010) and even taboo (Stone, 2013). It is life-changing for those left behind the deceased – it usually entails extra responsibility for them and can have a significant impact on their own well-being.

Funerals, therefore, represent potentially 'awkward' events for us – where awkward means 'not easy to handle or deal with; requiring great skill, ingenuity, or care' (Merriam-Webster Dictionary, 2020). Even if we hear that someone is attending a funeral, it can be difficult to know how to react or what to say; it can be similar if we are introduced to someone working in this industry. Despite this 'awkwardness', end-of-life events share the characteristics of many other events, where an event represents 'a unique moment in time celebrated with ceremony and ritual to satisfy specific needs' (Goldblatt, 1997, p. 2). As such, they represent significant and predictable opportunities for the end-of-life industry, which includes a range of event service providers who make notable contributions to successful outcomes, such as venues, caterers, flower arrangers and, more recently, even photographers and entertainers. Together with mourners, and potentially the deceased (in particular, if they have made their own end-of-life arrangements), these various stakeholders co-create the event.

Fundamentally, funerals represent a mechanism to help to maintain social order (O'Rourke, Spitzberg & Hannawa, 2011) and thereby support societal and community well-being. They offer re-assurance at a time of uncertainty; they maintain public health and safety, for example, reducing the spread of infection; and they offer the opportunity for 'sharing common experiences with one's fellows' or 'communitas' (Turner, 2012, p. 2). Funerals also make a significant economic contribution, with an even wider multiplier effect beyond the scope of this discussion. *The Telegraph* (the online version of *The Daily Telegraph* newspaper) quoted figures from the financial services company Sunlife which indicated that the total cost of dying in the UK added up to just under £10,000 in 2019 (Betts, 2020). This total included all professional services and 'discretionary extras' including the 'send-off', also known as the wake or more formally as the funeral reception. Based on the Competition and Markets Authority's (CMA, 2019) advice that, in 2017, 513,000 (almost 85%) of a total of 607,000 deaths involved a funeral, which suggests a total funeral value of almost £5bn. The CMA's report suggests that about a third of total costs related to flowers and catering (i.e., the more 'celebratory', events and venue-related dimensions of the occasion).

On a more personal level, the death of someone close to you is identified as one of the most stressful life events (Bisconti, Bergman & Boker, 2004; Sutin *et al.*, 2010). Despite the certainty of death, it still represents loss; even in the event of a 'good death' (Walters, 2004), we feel the deceased has been 'taken' from us. Feelings of stress and loss – along with depression, anxiety and the need for therapy – are linked to lower levels of subjective well-being (Steger *et al.*, 2006). Of the participants who reported specific personal events in Sutin *et al.*'s (2010) study, over one-third of the sample reported the death of a loved one as the most stressful event they had experienced in the last ten years. Furthermore, death often evokes acute reactions and disrupts normal self-regulation and equilibrium, notwithstanding that these outcomes can differ from person to person (Coifman, Bonanno & Rafaeli, 2007). Emphasising its transformative, negative effects, Steger *et al.* (2006, p. 90) propose that recovering from the death of a loved one can represent an 'existential crisis' (i.e., one which makes us question the meaning of our own existence (Cuncic, 2020)). Consequently, death is one of the most personally challenging phenomena we face in our lives, and it is no surprise that we may try to distance ourselves from it (Stone, 2013).

Funerals, and other end-of-life events or rituals, represent a significant intervention which can support both individual and group well-being. Like death itself, they are universal phenomena irrespective of cultures, spiritual beliefs and value systems (Hoy, 2013). Depending on context and perspective, funerals serve a number of different purposes, including assistance in the processing of grief; the opportunity for expressions of emotions for, and connections with, the deceased; and providing support for beliefs in spirit and the afterlife (O'Rourke, Spitzberg & Hannawa, 2011). Kisiara (1998, p. 127) summarises them as follows:

> the conduct of kin and non-kin participating in the funeral, expression of feelings and exchange of information among those attending the funeral, a

reminder to everyone of the inevitability of death and the need to prepare oneself in life, and affirmation and reaffirmation of the relationship between mourners, the deceased, and the bereaved family.

In Western tradition, funerals usually include a service (including a burial or cremation) which can be religious or not followed by a reception. For their own personal reasons, mourners may attend one or the other, or both. Both components (one typically more formal and one less formal, although this is changing) provide 'funeral-scapes' in which we can start to come to terms with loss and show/share our emotions. In terms of well-being, these emotional outcomes include grief, shock, anger, guilt and longing, and they can also be accompanied by physical reactions (Cruse Bereavement Care, 2020).

O'Rourke (2020) proposes that the existence of mourning rituals, albeit different across cultures, emphasises our humanity and the need for people to grieve together. Funerals, therefore, operate as a bonding mechanism, and it is not unknown for mourners to travel thousands of miles to attend a funeral in person (Holloway *et al.*, 2013) if the arrangements allow for it. This may explain our reactions to instances both when a funeral is not possible (perhaps in the absence of a body) or delayed and also 'notable deaths' which result in more public occasions. As an example of the latter, Haynes and Shearer (1998, p. 6) acknowledge 'an unprecedented outpouring of public emotion, not just in Britain but across the world' following the death of Princess Diana in 1997. Any absence of this mechanism, therefore, has potentially significant implications; for example, a situation in which the opportunity to grieve together is constrained – as we face at the moment due to the COVID-19 pandemic – could represent a successive, double blow to the well-being of the bereaved.

Fortunately for us, there is a relatively plentiful supply of places and organisations who choose to labour with death on a daily and regular basis by providing us with a range of related services. These groups include solicitors, stonemasons, counsellors, funeral directors, crematoria, and event venue and catering staff. Whilst there is a commercial aspect to the services provided, and notwithstanding concerns about the rising cost of funerals, there is evidence that such providers 'are obliged to charge for the goods and services they purvey, but they are largely motivated by care' (Bailey, 2010, p. 220). More recently, there has been a paradigm shift, with re-positioning of funerals as 'celebration of life' rather than 'end of life' events. Parsons (2012) notes that funerals have progressed from a predictable, equitably simple and formulaic structure to a personalised event, where loved ones can both say goodbye and celebrate the life of the deceased. Importantly, despite these changes, Holloway *et al.* (2013, p. 53) suggest that deep down at their very core funerals still remain about 'meaning-seeking, creating and taking', identifying that where meaning is lacking it contributes significantly to mourners' distress and therefore their well-being. This has significant implications for people working in the funeral industry, who need to be acutely aware that '[t]hese are special events with huge emotional significance and so subtle nuances and detail matter' (Laws & Deverell, 2019, p. 332).

A 2018 YouGov UK survey reveals that, whilst 42% of the adult population believes a 'proper send-off' is important, a greater proportion of 56% believe an important criterion is 'making it personal' (Assets.publishing.service.GOV.UK, 2019). 'Funeral-scapes' are also changing, with increasing acknowledgement of secular venues and accompaniments, and a move away from purely religious services (Laws & Deverell, 2019). In 2019, *The Telegraph* newspaper highlighted that sombre hymns are no longer the music of 'choice' at UK funerals, replaced instead by modern day 'pop' songs such as Frank Sinatra's *My Way*. Carson (2011) also acknowledges the introduction of 'living funerals', where the event acts as a pre-emptor to death, thereby enabling the significant person to be the guest of honour at their own end of life celebration. Given the aforementioned comments, this shift would seem to represent significant benefit to individual well-being as well as commercial opportunity, especially for funeral reception service providers such as venues and caterers. Admittedly, these trends may not represent the same opportunity for funeral service managers (such as funeral directors and arrangers) who may be working within stricter guidelines and especially in the case of religious service.

Conceptual framework

The purpose of this study is to consider roles, well-being and experiences within the context of funeral services and funeral receptions from the perspective of the event service providers. Event manager 'voices' are generally unheard in the events literature, and '[few studies] focus on event organisers, whose expertise may provide a better understanding of the connection between events' perceived impacts and well-being' (Slak Valek & Fotiadis, 2018, p. 41). Consequently, a sound conceptual framework and methodology for the study were required to make it as rigorous as possible. The framework used here specifically acknowledges *well-being* (individual and group), *co-creation*, *emotion management* and *tensions/conflicts*, which are now briefly explained.

Funerals represent significant *social occasions* (Kisiara, 1998; Laws & Deverell, 2019). Opportunities for social interaction are generally recognised as a source of positive feelings, and therefore positive impacts on, well-being – and especially related to feelings of companionship, 'communitas' and security during emotionally significant moments (Campos *et al.*, 2015).

Campos *et al.* (2015, p. 21) further identify that such social interactions represent co-created experiences, which are 'a matter of interpersonal, "outer interactions" with the experience environment, the experience people, and the experience activities'. Although the notion of co-creation has not been tested empirically in association with funerals in the events literature to date, it has been proposed explicitly by Laws and Deverell (2019) and acknowledged in funeral literature, for example:

> [Funerals are] "a collaboration of three, not two, parties: funeral professionals, close family members and the congregation as a whole."
>
> (Bailey & Walter, 2016, p. 164)

Whilst Holloway *et al.* (2013, p. 50) note in their investigation into the changing nature of the shape and content of funerals for the *Mortality* journal:

> It is this active participation by all actors in the co-creation and enacting of the funeral as a meaningful event which was at the heart of the funerals we studied.

As such, *co-creation* was identified as a significant lens through which to investigate the issues raised in this study.

Within the funeral context, Bailey (2010, p. 209) notes how funeral arrangers are tasked with balancing commerce and compassion in face-to-face provider-client encounters – 'working in a commercial relationship with highly vulnerable and emotional clients is central to the funeral director's job'. Whilst differences in duties vary, especially between different cultures and nationalities, Bailey's (2010) study identified four types of emotion management within the funeral arrangers' role: pecuniary (for commercial gain), prescriptive (according to professional standards), presentational (following social 'rules') and philanthropic (pastoral, including empathetic). Laws and Deverell (2019, p. 332) emphasise the increasing importance of the philanthropic dimension, quoting a funeral director's recommendation:

> If you're thinking about planning your funeral with a professional, find someone who understands the value of funerals and why we have them. Stay away from professionals who reduce funerals to three pieces of music and a pre-paid plan.

Drawing on Hochshild's seminal ideas introduced in *The Managed Heart*, Bailey (2010) also notes how various participants at a funeral may alter their inner feelings to align with what they consider to be the 'appropriate' emotion.

In order to deliver the required event outcomes, Phi, Dredge and Whitford (2014, p. 406) posit that it is important to appreciate potentially contestable issues to avoid what they call 'wicked' problems and to be especially sensitive if stakeholders' concerns are 'overlooked or minimised'. This notion would seem especially pertinent to the funeral context, since it is a 'once in a lifetime' gathering of people who are intimately and emotionally connected. Contestable issues and areas of potential conflict, therefore, represent obvious threats to the well-being of the stakeholders of funerals (Sofka, 2004; Woodthorpe, 2017). The Merriam-Webster Dictionary (2020) provides a useful, multidimensional definition of conflict for the purposes of this study, indicating that it can be something both 'internal' (within an individual, so e.g., within a mourner or service provider) and 'external' (between individuals, e.g., between mourners and/or as result of external factors).

Methodology

In keeping with the 'humanistic' nature of the study a constructivist (Denzin & Lincoln, 2000), triangulation approach comprising the collection and analysis of both primary and secondary data was adopted. Secondary research was collected from web-based resources, including 'grey literature' from government sites, charities and professional bodies; 'quality' newspapers; and academic books and journals. Given the lack of consideration of funerals in the events literature, journals such as *Death Studies* and *Mortality* provided a prolific source of insight. The UK National Association of Funeral Bodies was especially prolific in identifying areas of potential conflict amongst funeral stakeholders. As has already been highlighted, the events literature provided the main concepts and themes for the study.

There were various actual and perceived constraints on collecting primary data from funeral stakeholders during the recent COVID-19 pandemic. In particular, it was considered inappropriate to expect individuals already stretched, and potentially more stressed, in their work to discuss work during their 'down-time'. Therefore, primary data was restricted to three in-depth interviews. The participants represented funeral service managers (denoted as FSM1 and FSM2) and also one funeral reception manager (denoted as FRM1). All three participants were known to at least one of the authors. Online 'active interviews' were used to gather data. According to Holstein and Gubrium (1995), active interviews facilitate a process by which researchers and interviewees contribute more equally to the discussion – so essentially they have a 'conversation' – and thereby enable both to co-create meaning. They seemed especially relevant for the purposes of this study for a number of reasons, notably: the limited investigation of event managers' perspectives, especially using more insightful qualitative techniques (Slak Valek & Fotiadis, 2018); their use in similar types of study, including one previously produced by one of the authors (i.e., Fallon & Robinson, 2017); their acknowledgement that both interviewee and interviewer can contribute to a richer discussion (in this case, all participants were family with funeral services and receptions); and the aforementioned familiarity of the participants.

The interviews all took place following the introduction of the COVID-19 pandemic 'lockdown' restrictions, which enabled their impacts on funerals to be considered. They lasted between 40 and 60 minutes. Each interview was recorded and transcribed *verbatim* before the content of each one was then reviewed. The pre-identified themes were used to analyse the primary and secondary data. In some cases, these themes emerged 'naturally' – or unaided (Slak Valek & Fotiadis, 2018) – from the discussions. For example, the feelings related to, and concern for, well-being and the desire for a 'good send-off' were elicited freely from the interview conversations. In other cases, the themes were deliberately applied to the interview content, for example, the subject of tensions was generally avoided in the interview questions.

Findings and discussion

Within both the primary and secondary data, the themes or dimensions were often mentioned together. This is understandable given that they are inter-related, that is, the desire of the service providers is to co-produce a 'good send-off' and in doing so enhance the mourners' well-being, reduce stress and avoid conflict. As such, some of the quoted responses are 'cross-cutting'. In some cases, this is referred to more explicitly whilst in others it is left open to the reader's interpretation.

Well-being

Within the funeral service and reception contexts, two inter-related dimensions of well-being emerge. The overall or group dimension relates to the common goal, or main outcome, of creating a 'good send-off' for the deceased. The 'send-off' terminology is commonly used by different service providers, perhaps because of its euphemistic and inclusive nature. As such, it seems to indicate 'communitas' (Turner, 2012), as suggested by one of the funeral service managers:

> FSM1: 'When we are talking to our clients, they don't know if they are doing the right thing, there's often a lot of doubt, so we would re-assure them you are giving them a good send-off . . . it's a nice way of talking to our clients, it's a nice way for them to look at the service . . . for everyone to look at it.'

Similarly, the venue manager recognises the role of the reception, catering and associated staff in working towards a common purpose:

> FRM1: 'We are an important part of the send-off. You are there to provide support for them. I say that we provide 'tea and sympathy'. The main thing is listening and empathy, but the tea is the first thing people want when they come in.'

It is therefore important to avoid a 'bad send-off', and so be disrespectful:

> FRM1: 'So you make sure you don't run out of food – this is what people are afraid of. Then it's not a good send-off or sign of respect. It's showing that respect. . . . You also choose the right staff – some people cope, and some don't.'

This desire, or even pressure, from the living to give the deceased a good send-off is not restricted to 'nearest and dearest' and can come from other associated groups, as suggested by Covey (1989), for example, professional and community groups. In 2019, a Scottish armed forces charity ensured that ex-serviceman William Dunsmore, who died without any family at 80, 'will not make his final journey alone' by launching a 'heartbreaking' social media appeal (Stewart, 2019). As a result, 250 mourners attended his funeral. An SSAFA spokesperson commented that 'thanks to the fantastic response to our social media request I am sure his send-off did him proud'.

Supporting and respecting individuals, and/or individual groups, through the grieving process, as identified by O'Rourke, Spitzberg and Hannawa (2011), represent the second dimension of funeral well-being:

> FSM1: 'It [the send-off] helps people to acknowledge the loss . . . a lot of people when someone dies don't think it's real, they can go into a denial state. The service helps them with that process, it finalises it for them. In the grieving cycle, it helps them come to the acceptance stage.'

One funeral service manager describes the essential nature of her role, emphasising that it is about supporting the deceased as much as the mourners:

> FSM2: 'You have got to respect the deceased – that's someone's mum, sister, dad and you have to respect them as if they were still alive. . . . [In the Chapel of Rest] you are trying to make sure that they [the deceased] look their best so that they cause the least distress possible to their family visiting them.'

Similarly, the venue manager acknowledges that respect is equally due to the deceased and those grieving, and also their cultures and lifestyles:

> FRM1: 'Whoever the person was, they are still a person, aren't they? Just because they no longer exist, they still represent a person and you have to respect them. . . . It's also people's attitudes and beliefs, whether you agree or not you have to respect them, and not make value judgements.'

In supporting others, the service providers' own well-being can be affected, and as Steger *et al.*, (2006) suggest this can be in a positive and/or negative way:

> FRM1: 'You do get job satisfaction. People are usually so appreciative at the end of the event. It's the small things – they will put their hand on your arm when you are picking the plates. It brings a closeness – that's what you do it for.'
>
> FSM2: 'I love my job – there's no greater privilege than to be able to look after someone's loved one. I get a lot of satisfaction from it. People are kind – thank you cards are the main one.'
>
> FSM2: 'You are meant to be detached . . . sometimes it's harder than at other times. For example, with a quick succession of deaths, people have not recovered from one death and then are immediately thrown into the spiral again.'
>
> FSM2: 'Due to COVID-19, people have been limited in how they been able to grieve. That's been the most emotionally draining time of my career'

The last two quotes indicate emotional labour, demonstrating that alignment of feelings is not restricted to mourners (Bailey, 2010). FSM2 also identified that mourners are not exclusively human; it is not unknown for people visiting the deceased pre-funeral in the Chapel of Rest to bring the deceased's pets in with them, as a confirmation to the animals of their owner's death.

Co-creation

The 'good send-off' is co-created between the different stakeholders via both 'soft' skills, as previously identified by Bailey (2010) and Laws and Deverell (2019), and technical skills (e.g., from embalming to catering) and also tangible service elements (e.g., décor and food). Both the venue manager and the funeral managers emphasise that the key skill for service providers is empathy. Whilst empathy has commonly been acknowledged as a key service quality characteristic since the development of SERVQUAL (Parasuraman, Zeithaml & Berry, 1985), it is interesting to see here how all three respondents concur not only on its importance but also in how they create empathy, irrespective of their different roles in the send-off:

> FSM1: 'Empathy is really, really important, and also having an understanding. People can be angry about what's happened and the processes they now have to go through, and it's about understanding why they are angry.'
>
> FSM2: 'You have to be able to empathise with people. But you also have to be able to take yourself out of the situation they are in. If people are coming in absolutely bereft, there is no point in you getting upset at the same time, and sometimes it's easier to do than others. . . . You have to listen, you have to care. I am a caretaker for the deceased and their families. I am a listener . . . you have to be a listening ear.'
>
> FRM1: 'From a catering point of view [planning the reception], you are there to support them. The main thing is to listen to what they want and very often they will talk about the deceased. They want to talk to you and I really encourage that. Just listen and be interested, even if they have told you ten times.'

Covey (1989) suggests that the highest order form of listening is empathetic listening, which occurs when we move into the other person's frame of reference. All three respondents identify that their input is not restricted to listening, but generally being mindful, observant, compassionate, tactful and responsive, and not just directly with clients but also other mourners:

> FSM2: 'It's also observation . . . trying to cause the least distress.'
>
> FRM1: It's very emotional and if it's where alcohol is served [actual reception], sometimes people have too 'much to drink, so you can subtly mention it to one of the party that they may need some support. It's about being observant. The more experience you have, the more you can see things.'

The respondents also explain how they balance the frameworks within which their organisation operates, and the particular individual context and set of clients to work towards creating the best possible send-off, which is consistent with Bailey's (2010) identification of different dimensions of emotion management:

> FRM1: 'Very often it can be a blur – they are disoriented. They don't know all the details, it's short notice unless the death is expected. So you help them, you write everything down, offer them guidance on timings. . . . It's important to give them some control back. They feel better. People may not have a lot of money, but want something nice [for the send-off]. It may cause embarrassment, so have options to offer them. Make things simpler and cheaper for them, do 'specials' for them. They may ask 'Can we bring a cake?', and you can provide napkins for them. This also gives them input and control.'
>
> FSM1: 'We give them options, we never recommend . . . what's right for some families is not right for others, it's a personal preference. Sometimes within families there are different preferences, and emotions: "Why do I have to do this?" . . . "I don't understand." You have to stay calm. We can help them by providing options, such as taking care of some of the procedural things they have to do for them. We go the extra mile and we don't charge anything extra for that.'

Consequently, the family and friends of the deceased can be supported with the offer of more involvement and personalisation, for example, with photographs, music, speeches and readings, the choice of food and drink. The individuality of each occasion is further enhanced if the deceased can play a part in the co-creation, therefore, having 'their own way':

> FSM2: 'It's best to make plans, it's more common when the death is expected. . . . You wouldn't let someone else plan your wedding, so why would you let them plan your funeral!'

The co-creation can also be dynamic as the mood and flow of the event, and service provider roles, change (Getz, Andersson & Larson, 2007; Wallace & Michopoulou, 2019):

> FRM1: 'It may start off dark, but as it goes on you can hear people talking about instances that have happened, and memories . . . and people are laughing. And then they don't want to leave at the end.'

A more public and famous example of this fluidity was demonstrated at Princess Diana's funeral service at Westminster Abbey in 1997. The applause generated by those listening outside following the performances of Sir Elton John's *Candle in the Wind,* and Earl Spencer's tribute was subsequently taken up by the congregation inside Westminster Abbey. The example also emphasises differences between mourners, in this case, based on physical location, that is, outside or inside the service venue. It also suggests that location did not stop them from creating 'communitas' with each other, which has important implications for situations where mourners cannot physically all together in one place.

Funeral tensions

As indicated earlier, funerals are stressful and awkward occasions (Bisconti, Bergman & Boker, 2004; Sutin *et al.*, 2010). They can raise contestable issues that represent threats to both the send-off and the well-being of the stakeholders of funerals (Sofka, 2004; Woodthorpe, 2017). The findings so far have focused on the uplifting and bond-forming nature of services and receptions, notwithstanding the challenges that are faced by the main stakeholders. These bonds can also form – so be co-created – between service provider and mourner. Whilst the primary data was limited in generating issues related to the *tension* theme of this study, perhaps due to the sensitivities of the subject, it did identify some related issues.

All three respondents indicated that they operate very much with an empathetic, philanthropic philosophy rather than thinking more commercially, which supports the findings of Bailey (2010). This disposition may also reflect their willingness to participate in the study. Indeed, FRM1 emphasised that an awareness of the financial situation represents an important part of acting empathetically with those arranging the reception and thereby reducing embarrassment and awkwardness. She also identified that there are many unknowns, some of which differentiate funerals from other social events, for example, 'you don't always know how many people will turn up [at the reception]' and even if they turn up then they will not always want to eat. One of the funeral service managers expressed how important it is to manage the clients' expectations, which is common to all events but which is more sensitive here:

> FSM1: 'It's not good if you make a promise and it doesn't happen. It's really about managing expectations in a sensitive manner. We give them options, we never recommend . . . what's right for some families is not right for others, it's a personal preference. Sometimes within families there are different preferences.'

FSM2 also noted other 'unknowns', ranging from who wants to visit the deceased in the Chapel of Rest and whether individual mourners will attend the funeral on the day to the weather on the actual day. Whilst she was aware of conflict within families, for example, via a request to bar individuals from seeing the deceased prior to the funeral, she had also heard anecdotally about:

> FSM2: 'fights in the crematorium. It's a very emotional time, especially when you've got fractured families. Funerals are tinderboxes of emotion . . . like weddings.'

This suggests that empathy includes awareness of sensitivities such as family conflicts or differences, that is, the 'nuances of the situation' (Laws and Deverell

2019). Similarly, FRM1 notes that 'in some cases nobody wants to take charge, in others everybody wants their say' in making the reception arrangements, so she acts as a kind of 'peacemaker' to help agreement to be reached. Phi, Dredge and Whitford (2014) note that acting inclusively with stakeholders represents an important skill for event managers generally. As already suggested, if the deceased make their own plans, or at least makes their wishes known, this can help to reduce tensions.

The secondary data provides greater acknowledgement of more serious conflicts that are co-created within the context of end-of-life events. A survey amongst its four thousand UK funeral homes members by The National Association of Funeral Directors (NAFD, 2017) noted an increase of almost 60% in family conflict 'which had led to funeral directors being targeted and in some cases having to take extreme action to make peace among warring families'. The survey identified the three commonest sources of conflict as the coming together of estranged families, uncertainty of the deceased's wishes and lack of provision of payment for the arrangements. NAFD's President at the time suggested that a lack of pre-planning by the deceased:

> is leaving families with uncertainty that is increasingly exacerbating fault lines and turning into conflict. As well as having a terrible impact on families, this is also exposing funeral directors and their teams to aggressive behaviour and increasingly leaving them with unpaid debts too.

Whilst these considerations generally suggest that funerals represent broadly predictable but evolving service encounters, recent events at the time of writing demonstrate that nothing can be taken for granted. The COVID-19 pandemic of 2020 and subsequent reactions – especially by the government – have had a devastating impact on the holding of events in general and on the nature of funerals in particular. The UK government has changed its guidance on holding funeral services and receptions, limiting the number of attendees to 30 and keeping to social distancing rules (GOV.UK, 2020). If the virus was suspected as the cause of death, the requirements are even more severe, for example, limiting embalming and viewing the deceased. The significant role of the government here also re-emphasises the need to acknowledge the fluidity of event stakeholder roles (Getz, Andersson & Larson, 2007; Wallace & Michopoulou, 2019). Whilst it is beyond the scope of this discussion to consider the potentially significant and wide-ranging impacts of the virus and subsequent restrictions, the respondents both noted that from their perspective it had a negative effect on both the send-off and the well-being of individuals:

> FSM1: 'It's made it very difficult, especially where the client hasn't seen the deceased for a long time even before their passing. I felt especially bad as families haven't got what they should have got from the funeral . . . and as the regulations have relaxed over time, you know that some people have had far less options than others earlier on.'

> FSM2: 'Due to COVID-19, people have been limited in how they been able to grieve. You couldn't give them a hug. That's been the most emotionally draining time of my career.'
>
> FSM2: 'COVID-19 has limited in how they have been able to grieve. It's prevented the full celebration of life. In some cases, people have not been able to visit them before they died. It has been emotionally taxing to watch family.'

Given the preceding disclosures around well-being, it is not surprising that, irrespective of their different roles, all three interviewees have personally struggled here. The subsequent restrictions have had a profound impact upon not just on deceased's families and their grieving but also on the managers' emotion management and *modus operandi*. As Bailey (2010) suggests, not only admittedly in the context of mourners but also potentially more widely applicable, there are longer-term implications if grief and/or relief are subsumed for the 'benefit of the occasion'. Perhaps, this is also missing from wider and contemporary public consideration of the impacts of the curveball that is COVID-19? If so, it is especially pleasing to have this opportunity to give a 'voice' here to a group of service providers who in their own way provide a vital and socially significant 'emergency service' with such high levels of passion and compassion.

Conclusions

This chapter sheds light on a dimension of the events industry that has so far remained largely 'in the dark' academically speaking. Given the awkwardness associated with death, this oversight is understandable. The rationale for this intervention includes, inter alia, changes in funeral trends and attitudes towards end-of-life events; the commercial and career opportunities offered to existing event venues and managers; the need for an increased understanding of the significant potential impacts of phenomena such as COVID-19 on the funeral sector; and the need for an improved understanding of the skillset needed to work in specific events contexts and, in particular, those related to funeral receptions.

The first objective of the study was to give a 'voice' to end-of-life event managers to understand the nuances of their work and the challenges they face; it is also linked to the second objective in terms of contextualising their work within broader events practices. Via a triangulation of interviews and a review of secondary data, there are some key messages to be shared. The findings indicate that there are both similarities and differences with the general events industry and also between funeral service and funeral reception managers. General event themes related to well-being, co-creation, stakeholders and tensions are applicable to end-of-life events as much as other events. Like other events, funeral experiences are co-created between various stakeholders, and managers can make significant interventions into securing a successful outcome and a general sense of well-being around the event. The most similar related event is arguably the wedding (Laws & Deverell, 2019). However, funerals also differ from other events, especially due to: their primary purpose of grieving; an absence of planning; their short-term nature; and high levels of 'unknowns' and stress. Given these characteristics, and as in other event contexts, an awareness of ideal skills to work on either funeral services or receptions offers the event business the opportunity not only to differentiate itself but also to preserve, and even enhance, the well-being and work satisfaction levels of its staff. These skills are largely 'soft' ones relating to empathy and awareness (especially relating to the nuances of the specific situation) and are shared between different service providers. Whilst these skills may represent a particular disposition, FRM1 also considers that this can be learned, and all three respondents suggest that the skill level develops over time. These skills can therefore be invaluable in helping the service provider not only to initially cope with grieving, disoriented and angry clients but also to support and guide in the design of a 'fitting send-off' and its related outcomes.

In order to satisfy the second objective, the authors have developed the CRADLES 'Soft' Skillset for Funeral Service Providers (Figure 8.1) to emphasise the ideal emotional intelligence dimensions for working in the funeral industry and/or on funeral-related events. It offers a valuable breakdown of skills that could be developed by event/venue managers who are looking to develop the opportunities offered by end-of-life events.

Parasuraman, Zeithaml and Berry (1985) famously identified four skills within their five-dimensional SERVQUAL model, that is, Reliability, Assurance, Empathy and Responsiveness. These dimensions have been identified in a wide range of service contexts. More specifically, the CRADLES framework emphasises and further illuminates the 'Empathy' dimension, reflecting the unique, sensitive and emotional nature of the end-of-life context. The primary and secondary data indicate that certain behaviours and practices – in particular, those related to respect, listening, compassion, awareness, tact and diplomacy, calmness and

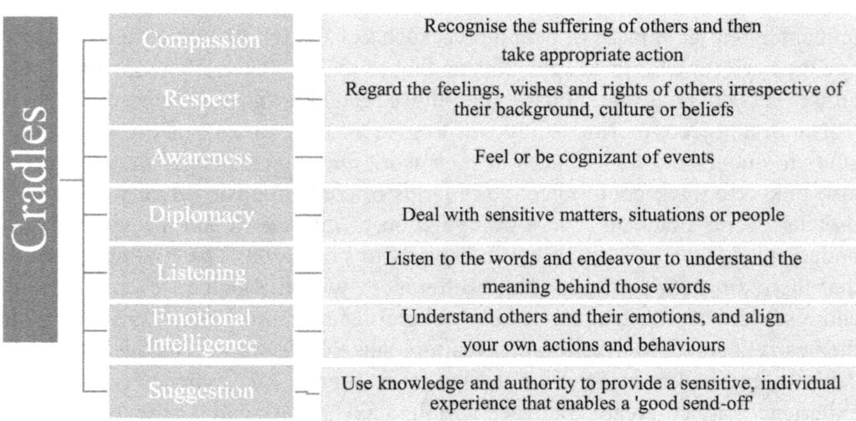

Figure 8.1 The cradles 'soft' skillset for funeral service providers

Source: Authors.

suggestion (including creativity) – contribute to both a fitting and personalised send-off which can be more harmonious, inclusive and uplifting for all concerned. The model also potentially provides a useful framework for both recruitment and personal risk assessment, for example, with the intention to improve positive and reduce negative well-being of those working in end-of-life settings.

The final aim of the chapter was essentially to raise the profile of funerals for the events industry and its stakeholders (especially, businesses, venues, teachers and students). In terms of scale, the end-of-life statistics 'speak for themselves', and, if anything, the COVID pandemic has added to these. Hopefully, a further legacy of the pandemic will be an enhanced understanding of the significant role of the events industry, and those working in it, in making sense of the relationship between life and death. Although not acknowledged formally in this study, there are commercial benefits to be gained here, with high performances potentially leading to positive 'word of mouth' and even repeat patronage by satisfied clients.

References

Bailey, T. (2010). When commerce meets care: Emotion management in UK funeral direction. *Mortality*. 15 (3), 205–222.

Bailey, T., & Walter, T. (2016). Funerals against death. *Mortality*. 21 (2), 149–166.

Betts, H. (2020, February 1). The cost of dying is now £9,493. . . So is it time to shake up the overpriced funeral industry? *The Telegraph*. Available from www.telegraph.co.uk/family/life/cost-dying-now-9493-time-shake-overpriced-funeral-industry/ [Accessed 1 September 2020].

Bisconti, T.L., Bergman, C.S., & Boker, S.M. (2004). Emotional well-being in recently bereaved widows: A dynamic systems approach. *The Journals of Gerontology: Series B: Psychological Sciences and Social Sciences*. 59B (4), 158–168.

Campos, A.C., Mendes, J., Oom do Valle, P., & Scott, N. (2015). Co-creation of tourist experiences: A literature review. *Current Issues in Tourism*. 21 (4), 369–400.

Carson, D. (2011). *Parting ways*. University of California Press. Available from www.ucpress.edu/book/9780520268739/parting-ways [Accessed 1 September 2020].

CMA. (2019, 28 March). Competition and markets authority. *Funerals Markets Study*.

Coifman, K.G., Bonanno, G.A., & Rafaeli, E. (2007). Affect dynamics, bereavement and resilience to loss. *Journal of Happiness Studies*. 8, 371–392.

Covey, S. (1989). *The seven habits of highly effective people*. London: Simon and Schuster UK Ltd.

Cruse Bereavement Care. (2020). You are not alone. We're here to support you. Available from www.cruse.org.uk/ [Accessed 13 August 2020].

Cuncic, A. (2020). What is an existential crisis? *Very Well Mind*. Available from www.verywellmind.com/coping-with-existential-anxiety-4163485 [Accessed 1 September 2020].

Denzin, N.K., & Lincoln, Y. (2000). *Qualitative research*. Thousand Oaks, CA: Sage.

Fallon, P., & Robinson, P. (2017). 'Lest we forget': A veteran and son share a warfare tourism heritage experience. *Journal of Heritage Tourism: Children, Families & Heritage*. 12 (1), 21–35.

Getz, Dhersson, T., & Larson, M. (2007). Festival stakeholder roles: Concepts and case studies. *Event Management*. 10 (2), 103–122.

Goldblatt J.J. (1997). *Special events: Best practices in event management*, 2nd ed. New York: VNR.

GOV.UK. (2019). *State of the nation 2018–10: Social mobility in Great Britain*. Paper presented to Parliament pursuant to section 8B(6) of the Life Chances Act. Available from www.google.com/url?sa=t&rct=j&q=&esrc=s&source=web&cd=&ved=2ahUKEwjb9u33-8jrAhXLbsAKHezUCrIQFjAAegQIBhAB&url=https%3A%2F%2Fassets.publishing.service.gov.uk%2Fgovernment%2Fuploads%2Fsystem%2Fuploads%2Fattachment_data%2Ffile%2F798404%2FSMC_State_of_the_Nation_Report_2018-19.pdf&usg=AOvVaw2BjR-nnDhThnnHIKCNDvJe [Accessed 1 September 2020].

GOV.UK. (2020). Arrange the funeral. Available from www.gov.uk/after-a-death/arrange-the-funeral [Accessed 5 August 2020].

Haynes, J., & Shearer, A. (1998). *When a princess dies*. London: C.G. Jung Analytical Psychology Club.

Holloway, M., Adamson, S., Argyrou, V., Draper, P., & Mariau, D. (2013). Funerals aren't nice but it couldn't have been nicer: The makings of a good funeral. *Mortality*. 18 (1), 30–53.

Holstein, J.A., & Gubrium, J.F. (1995). *The active interview*. Thousand Oaks, CA, London: Sage.

Hoy, W.G. (2013). *Do funerals matter? The purposes of practices of death rituals in global perspective*. Hove, England: Routledge.

Kisiara, R. (1998). Some sociopolitical aspects of Luo funerals. *Anthropos*. 93 (1/3), 127–136.

Laws, C., & Deverell, K. (2019). Events management for the end of life – mortality, mourning and marginalisation. In: Walters, T., & Jepson, A. (eds.) *Marginalisation and events*. Abingdon: Routledge, pp. 222–241.

Merriam-Webster. (2020). Available from www.merriam-webster.com/dictionary/conflict [Accessed 7 August 2020].

NAFD.org.uk. (2017). Family conflict over funeral arrangements up by almost 60% | *National Association of Funeral Directors*. Available from https://nafd.org.uk/2017/05/15/family-conflict-funeral-arrangements-almost-60/ [Accessed 6 August 2020].

O'Rourke, M. (2020). We're used to grieving together. What happens when we can't?: Death without funerals will be harder to adjust to than mere social distancing. *Washington Post*. Available from www.google.com/amp/s/www.washingtonpost.com/outlook/funerals-coronavirus-grieving/2020/03/20/bc07aafc-69f3-11ea-abef-020f086a3fab_story.html%3FoutputType%3Damp [Accessed 1 September 2020].

O'Rourke, T., Spitzberg, B.H., & Hannawa, A.F. (2011). The good funeral: Toward an understanding of funeral participation and satisfaction. *Death Studies*. 35, 729–750.

Parasuraman, A., Zeithaml, V.A., & Berry, L.L. (1985). A conceptual model of service quality and its implications for future research. *Journal of Marketing*. 49 (Fall), 41–50.

Parsons, B. (2012). Identifying key changes: The progress of cremation and its influence on music at funerals in England, 1874–2010. *Mortality*. 17 (2), 130–144.

Phi, G., Dredge, D., & Whitford, M. (2014). Understanding conflicting perspectives in event planning and management using Q method. *Tourism Management*. 40, 406–415.

Slak Valek, N., & Fotiadis, A. (2018). Happiness as a value of event organizers in Abu Dhabi. *International Journal of Event and Festival Management*. 10 (1), 34–47.

Sofka, C.J. (2004). What kind of funeral? Identifying and resolving family conflicts. *Generations*. 2 (5), 21–25.

Steger, M.F., Frazier, P., Shigehoro, O., & Kaler, M. (2006). The meaning in life questionnaire: Assessing the presence of and search for meaning in life. *Journal of Counseling Psychology*. 53 (1), 80–93.

Stewart, S. (2019). Hundreds attend old Glasgow soldier's funeral after heartbreaking social media appeal. *Glasgow Live*. Available from www.glasgowlive.co.uk/news/hundreds-attend-old-glasgow-soldiers-16427661 [Accessed 1 September 2020].

Stone, P. (2013). Dark tourism scholarship: A critical review. *International Journal of Culture, Tourism and Hospitality Research*. 7 (3), 307–318.

Sutin, A.R., Costa, P.T., Jr., Wethington, E., & Eaton, W. (2010). Turning points and lessons learned: Stressful life events and personality trait development across middle adulthood. *Psychology and Aging*. 25 (3), 524–533.

Turner, E. (2012). *Communitas*. New York: Palgrave Macmillan.

Wallace, K.,& Michopoulou, E.E.(2019). The stakeholder sandwich: A new stakeholder analysis model for events and festivals. *Event Management*. 23 (4), 541–558.

Walters, G. (2004). Is there such a thing as a good death? *Palliative Medicine*. 18 (5), 404–408.

Woodthorpe, K. (2017). Family and funerals: Taking a relational perspective. *Death Studies*. 41 (9), 592–601.

9 The lure of vinyl

Spinning 'open decks' events for well-being outcomes

Alex W. Grebenar

Introduction

> *The lure of vinyl. The feel of the sleeve. The bump of the needle. The crackle of the groove. And the music begins.*

The love of music, and specifically popular music, has endured and blossomed in every generation in the century or so since it became commercially available, immortalised on record. Indeed, in this time, music consumption has been shown to engender positive psychological outcomes for the individual. This could be the rush fuelled by endorphins (or indeed other chemicals) expressed best through dancing; according to Chic in their hit 'Everybody Dance', it can even 'make you happy again' (Rodgers & Edwards, 1979), a process borne out on millions of dance floors through time. It could be thrill of fandom, or it could be strong experiences of music (SEMs) which provoke deep emotional responses. It could be the lure of a 'scene' enticing devotees of their time, whether to the coffee bar culture of the 1950s and 1960s (Haslam, 2016) or the house music boom of recent years (Grebenar, 2020). Similarly, the simple pleasures of consuming music, socialising, experiencing the buzz of the new the charm of the old provide a lens through which we may understand the positive and rewarding outcomes of music events. Testament to this is the highly developed music infrastructure and 'ecology' across the UK (van der Hoevens & Hitters, 2019) and particularly in large cities, meaning that live music experiences are readily available to much of the population. These venues have a strong lineage dating back to the music hall, with changes, developments and trends dictating how music venues have adapted or died (Haslam, 2016). Events studies now form a well-established research field, with a large quantity of literature exploring their social role (Deery & Jago, 2010; Richards, de Brito & Wilks, 2013; Getz & Page, 2019; Devine, Quinn & Devine, 2020), and the somewhat elusive concept of 'experience' (Petterson & Getz, 2009; Grebenar, 2020). This research seeks to contribute to current understanding by developing a novel perspective to consider the vinyl record as a catalyst for well-being through the eventscape.

The link between events and well-being is now established with a not only burgeoning literature covering a range of issues as an ongoing stimulus of well-being

(Packer & Ballantyne, 2011) but also considering events as interventions to purposefully stimulate well-being outcomes for specific groups (McDermott, Orrell & Ridder, 2014). There is also some existing literature on the impact of music events, though much tends towards the music, rather than the eventscape, as the key element. This research therefore seeks, by way of a multidisciplinary approach, to develop an understanding of the well-being outcomes from music events – specifically within the context of one of music's most enduring and recognisable artefacts – the vinyl record. Presumed dead as a commercial product as recently as ten years ago, the vinyl record has proved to be one of the few success stories in the record industry's struggles in the era of music digitisation (Palm, 2017). Whilst the overall physical sales have declined, replaced overwhelmingly with streaming and downloads (Palm, 2017), vinyl sales have grown hugely since 2010, accounting for 1 in 8 of albums sold in the UK in 2019 (Helfet, 2020). Conspicuous consumption and imagery of the record contrast with the formless and furtive nature of songs played via streaming platforms. The intention of this research is not to proffer the superiority of vinyl as a medium *per se* but rather to distinguish the role which this physicality has on the experience of music events. Given this paradoxical rise in popularity in the face of technological innovation which dominates current social discourses (Palm, 2017), vinyl-only music events are an interesting by-product of this popularity. One particular event type, the 'open decks' event, forms the focus of this research; this is a relatively new phenomenon whereby participants may attend the event with their own records which they are permitted to play on supplied equipment in similarity to the well-established 'open mic' format for musicians (Behr, 2012). Whilst this was a necessary component of events in previous decades such as school discos whereby children would bring their own records to be played, the modern iteration is more performative in nature and thus can be seen in the context of the vinyl's reaffirmed position as a commercial product. Further explanation of the open decks event is provided in the Findings and Discussion section; it is believed that open decks events are introduced to the event studies literature for the first time in this chapter. Currently, there is some tangential research discussing the link between vinyl records and well-being, though none with this direct focus of the eventscape.

The aim of this research is to observe and conceptualise the impacts which open decks events have on the subjective well-being of the participating individuals. This is achieved by answering the following questions:

1 How are open decks events performed?
2 What well-being outcomes are achieved through such events?
3 What role do vinyl records play in those well-being outcomes?

Literature review

The re-emergence of vinyl as a viable commercial product is one of the more notable developments for the music industry in the 21st century – though in fact vinyl never really went away (Palm, 2017). Indeed, though vinyl sales dropped precipitously to the point at which mainstream retailers moved decisively to CD (Plasketes, 1992; Hayes, 2006), a vibrant second-hand and specialist subculture, characterised through markets and independent record shops, endured and now reaps the rewards. Sales of vinyl in UK and US markets are now at comparatively record highs and appear secure for the time being; however, it can also be noted that the consumption of vinyl in 2020 is radically different from that in, say, 1970, which could be described as vinyl's heyday (e.g. Hayes, 2006). As such, this warrants new research to explore how it can be experienced in the modern eventscape.

This difference has been noted in academic and commercial research. Much of vinyl's present iconography acts as some form of rejection of modernity and modern technology (Hayes, 2006; Lamont, 2011; Bates, 2019), with what was seen formerly as an obsolete technology reviving the tactile relationship which was a necessity in the pre-digital music landscape. This is an interesting paradox because despite this sense of authenticity, vinyl production/consumption is just as reliant on the economics of the music industry as any other product (Plasketes, 1992; Hayes, 2006). Furthermore, vinyl is comparatively expensive not only compared to the market price in past decades but also compared to alternative formats including CD, download and streaming subscriptions. This casts vinyl as a premium 'lifestyle' product – indeed, a notable portion of this market is dedicated to (sometimes spurious) reissues and sprawling luxury box sets (Zadeh, 2014), which often retail for over £50 and even £100+. Even at the opposite end of the market, a new vinyl record from a band with limited national profile will retail in the region of £20. This brings marketing connotations; on the one hand, vinyl not only can be luxurious, valuable and important but can also be exploitative. Indeed, the prescient lyric '*reissue, repackage, repackage*', from 'Paint a Vulgar Picture' by The Smiths (Morrissey, 1987) has become something of a truism, often repeated by cynical commentators on release of the latest overpriced or superfluous reissue (Haynes, 2017). Further evidence of vinyl's re-emergence as an iconic cultural artefact is the worldwide Record Store Day (RSD), ostensibly designed to stimulate demand for independent record stores through unique releases and local events. There are, however, multiple discourses at play (Bates, 2019); there is the accusation that RSD privileges bigger and major record labels, particularly where the rush to create special releases causes bottlenecks at pressing plants in which smaller labels are far down the pecking order. Nonetheless, RSD has revalorised the vinyl and created its own 'national holiday', further cementing the consumption of records within modern consumer culture.

Taking a more philosophical approach, we may also note the difference in the semiotic meaning of vinyl in 2020. Bartmanski and Woodward (2015, p. 4) note the 'symbolic narratives' now attached to its value in a sort of praxis brought on through vinyl's re-emergence. Whereas vinyl was once the only reliable method of consuming music on-demand (Plasketes, 1992), it is now merely one option, several of which are free-to-use. As such, the active choice in preferring vinyl to other (i.e. digital) forms of music is a notable signifier of personality and lifestyle. These signifiers have been variously noted in the literature; Lamont (2011) notes a rejection of the faster pace of life brought on by advances in modern technology, which vinyl embodies through a more involved and intricate playing ritual (Hayes, 2006; Behr, 2012). This also relates to the concept of emotional resonance from materiality (Bennett & Rogers, 2016; Wood & Moss, 2015), which is absent from digital formats, a resonance that Bates (2019) terms 'emotional-affective attachment'. It confounds the notion that technology is linear and irreversible in its progress (e.g. Plasketes, 1992; Bartmanski & Woodward, 2015) by placing emphasis on the medium as well as the music.

It can be argued as a postmodern *reification* of music – a re-emergence of music as a tangible entity whereby the materiality (Bates, 2019) becomes an important signifier of the power of music (which is by nature intangible). It is not just the physical manifestation of music *per se* but also the very nature of vinyl (Plasketes, 1992; Bartmanski & Woodward, 2015; Bates, 2019). Vinyl, through its signifiers, may give the user a greater sense of authenticity in consumption of music (O'Grady, 2015). This is noted specifically by Bates (2019), with flaws and surface noise now seen as an authentic and almost desirable element of vinyl, whereas digital music was sold to eliminate this nuisance (Plasketes, 1992; Hayes, 2006). Indeed, it is this concept of 'value' which has now shifted; Frith (2007) notes that from the late 1990s and the advent of internet music consumption, the value of recorded music (in both financial and philosophical terms) has declined. This fall in financial value may account for the higher market price of vinyl as labels and artists now rely on sales of physical merchandise more than ever (Bates, 2019;). This is an ontological multiplicity (Bates, 2019); vinyl is simultaneously old yet novel, pure yet mass-produced, authentic yet often just as corporate as any other product (especially major label releases). As such, it retains an elusive allure explored in detail in this research.

In recent years, the beneficial effects of events have been subjected to increased interest and research, leading to a better understanding of the role of events in personal and subjective well-being (e.g. Packer & Ballantyne, 2011; Weinberg & Joseph, 2017; Jepson, Stadler & Spencer, 2019). There is an overwhelming consensus that event participation can have direct and positive effects on well-being (e.g. Clift & Hancox, 2010; Bosnjak *et al.*, 2016; Armbrecht & Andersson, 2019; Jepson, Stadler & Spencer, 2019), and participation in social events has become an established part of interventions for these reasons and more (e.g. Hacking *et al.*, 2006; McDermott, Orrell & Ridder, 2014). Events are noted as drawing out specific emotional responses (Lamont, 2011; Wood & Moss, 2015; Armbrecht & Andersson, 2019); indeed, Weinberg and Joseph (2017) note that well-being outcomes can be a key motivator for attending an event (also Packer & Ballantyne, 2011). Whilst there are several types of well-being as defined through psychology and physiology (Haley & Senior, 2007; Henry, 2007; Packer & Ballantyne, 2011; Lamont, 2011; Stebbins, 2018), the key measurement at the intersection of events and well-being appears to be subjective well-being (Packer & Ballantyne, 2011; Lamont, 2011; Jepson, Stadler & Spencer, 2019). This is a measure of how the individual feels as a 'multidimensional construct' (Henry, 2007, p. 27), rather than of an objective wellness in terms of physical or mental health (Boer & Fischer, 2010; Yolal *et al.*, 2016). This subjective measurement appears ideal for phenomenological research within the event studies field and thus has been included as the focus of this research.

Events can stimulate a spectrum of feelings, and much music events research has taken this perspective. Chief amongst these are the 'peak experiences' (Gabrielsson, 2002) and 'strong experiences of music' (SEMs) (Lamont, 2011), which are experienced in the context of deeper meaning, a concept crucial to a sense of subjective well-being (Boer & Fischer, 2010; Lamont, 2011; Kawase & Obata, 2016; Stebbins, 2018). Lamont (2011) explores the concept of 'eudaemonia', the sense of living life in an authentic or satisfying way (see also Stebbins, 2018), and this relates to other studies that show music events as a way of expressing identity (Grebenar, 2020), feelings (Haley & Senior, 2007; Henry, 2007; Wood & Moss, 2015; Kawase & Obata, 2016) or rejection of certain social norms (Hemmett, 1998; Hesmondhalgh, 2000; O'Grady, 2015). The potential for deep emotional immersion has also been noted as a means to subjective well-being through participation in music events and festivals (Packer & Ballantyne, 2011). This may even extend to 'serious leisure' (Stebbins, 2018), where leisure experiences become highly substantial and conducive to the sense of self and subjective well-being.

Further to this established link between attendance at music events and well-being, there is also a notable consensus on the positive role of art participation, and specifically music performance, in a number of well-being measures (Haley & Senior, 2007; Sixsmith & Boneham, 2007; Clift & Hancox, 2010; McDermott, Orrell & Ridder, 2014; Bosnjak *et al.*, 2016; Stebbins, 2018; Vougioukalou *et al.*, 2019). Crucially to the originality of this research is that the existing literature appears to focus only on 'primary performance' (Frith, 2007) – where the individual creates something original, compared to 'secondary performance' which is the playback of existing material. Nonetheless, by analysing existing discourse, we may observe this eudaemonic outcome in the 'secondary performance' of open decks events through specific similarities with primary performance. Whilst purists may object that this is not the true essence of 'performance', this ignores the trajectory of the DJ into a hybrid performer of status (Haslam, 2016; Grebenar, 2020) and also ignores the potential for public 'performance' in its most literal sense to be subsumed into our understanding of musical performance.

Far from being a passive act, selecting and playing records to an audience require skills like any other type of performance (Behr, 2012). Firstly, there is the ability to operate the mediating technology; primarily not only a record player (often a double deck) but also the amplification system. This is rarely acquired by formal means (e.g. Behr, 2012) and thus requires experience or knowledge gleaned from others (potentially from a similar event and the opportunities they provide – Sixsmith & Boneham, 2007). Fundamental to this is the record player itself; most of the free-arm players require precision and delicate motion to ensure no damage occurs to the record or stylus. Beyond this, there is the potential for extra performance elements such as cutting, mixing and beat-matching which require timing and practice. They are also improvisational, again requiring specific skill and understanding to execute effectively. This could not realistically be achieved by a beginner, just as a novice would struggle to perform a song on a musical instrument. Beyond this technical aptitude which can engender positive well-being outcomes (Bosnjak *et al.*, 2016), there is also the need for sufficient cultural/subcultural capital (Thornton, 1995; Grebenar, 2020) in selecting and playing appropriate music at an appropriate time. Considering that a record collection is a discrete set of objects and thus finite in its reach, this arguably ascribes a sense of value and ownership to the performance, and records, in the same way in which a musician is bound by the limits of their ability and stagecraft (Behr, 2012). As such, this research suggests that to better understand the role of music performance on well-being, we must challenge the assumptions around what constitutes musical performance. The re-emergence of vinyl as both a musical medium and a performative artefact supplements this view, and so within the context of this eventscape the polysemy of vinyl, and the vinyl DJ, is crucial.

In conclusion, it can be observed that, within musical performance, well-being outcomes are bound up not just in the music itself but also perceptions of aptitude, performance and indeed the means through which the music is conveyed to the audience (and indeed its relationship with the audience). This goes some way to explaining the draw of vinyl whether as a novel curio re-emergent in a digital age, or as an enduring edifice of recorded music; as such, this chapter now turns to primary research as a means of exploring this in the context of subjective well-being.

Methodology

This research, conducted in summer 2020, takes a constructivist approach, assuming that knowledge is not objective but necessarily subject to the processes of the human mind; therefore, this research is positioned as an exploration of individual experience. The research utilises the phenomenological epistemology, foregrounding the experience and worldview of the research participants within the outcomes of the study. The research is qualitative and focuses on thoughts and feelings through semi-structured interviews with three event managers/DJs who currently run vinyl-only open decks events. The interviews were between 30 minutes and 1 hour in length, using in-depth questioning allowing for probing. In relation to the research questions, the interview schedule included questions on:

- The nature of the event in question
- Issues of well-being
- The role of vinyl records in the event and its well-being outcomes

Preliminary research identified 12 UK vinyl-only open decks events (not intended as an exhaustive illustration of prevalence), discounting any which also allowed digital files (such as DJ mix showcases). Of the individuals contacted by using a purposive selection technique, five responded, with three ultimately interviewed. Although each event is different in its own way, they are not used as case studies. All three interviewees participated in accordance with ethical guidelines with informed consent and the right to withdraw. All agreed to be named and are listed in Table 9.1 with some information about the respective events.

Table 9.1 Interviewee and event details

Interviewee	*Location*	*Event Name*	*Event Nature*
Joe Florek	Chorlton, Manchester	*Our Vinyl Weighs A Ton (1)*	Monthly; five 7″ records or three 12″ records
Stephen Evett	Shrewsbury	*Bring Your Own Vinyl (2)*	Weekly; timed slots
Michael Hodgson	Clapton, London	*Super Sevens (3)*	Fortnightly; timed slots, 7″ records only

Due to the closure of the UK venues during the COVID-19 pandemic, events 1 and 2 did not take place during the time of research, however, event 3 switched to online delivery at this time in otherwise the same format. Two interviews were conducted via Zoom, and one interview was conducted via telephone. The interviews were transcribed and subject to thematic analysis. The analysis identified 23 individual themes, which were coded accordingly and later combined to produce the research outcomes.

Findings and discussion

How are open decks events performed?

The 'open decks' as a participatory event appears to be a recent creation, a supposition supported by all three interviewees. JF noted the similarity to the school disco of the pre-digital era whereby children brought records to be played at the disco, presaging the 'user-generated' nature of open decks in a different environment. This tradition presumably ended sometime around the late 1980s/early 1990s coinciding with the decline of vinyl ownership. All interviewees believe that 'open decks' is a recent event invention in the time period coinciding with the resurgence of vinyl in the UK. SE suggested that this is due to lack of a 'home buyer' market for vinyl in the intervening years; whilst professional DJs may have bought and played contemporary vinyl, the wider public would be restricted to older vinyl which at that time may have become seen as lacking value or interest (also JF).

The format of the open decks event is relatively fluid, unlike other participatory events such as open mic nights (Behr, 2012), and all three events were different from each other (as shown in Table 9.1). All interviewees also reported subtle changes to the format over time. However, whilst the format is fluid, the basic concept of unpaid performers who play their own records to the audience remains the basic principle. Furthermore, the philosophical 'performance' of the events was remarkably similar in terms of approach, actions and outcomes. As such, these sub-themes have been grouped into two elements: the 'performative' and 'community' elements, as will now be discussed in turn.

Performative elements: Key to the performance of open decks events is the presence of specific rules of engagement – though MH characterised these as '*focus*', rather than exclusionary devices. Nonetheless, the environment which privileges vinyl becomes a central part of the experience. MH even noted a jovial 'militant' nature, whereby the event community 'polices' the rules relatively strictly, whilst also bending them to occasionally allow 10″ records. This suggests a performative element whereby the privileging of vinyl creates a specific arena in which to consume and contextualise music; it also, as noted by MH, gives the event potential '*cut-through*' in the crowded night-time economy of London. Indeed, whilst none of the three events had a specific music policy which included/excluded certain genres, there was an implicit assumption that obscure or less commercial choices were actively encouraged and elicited greater 'cache' (JF). This was observed by JF at other events; he noted that certain open decks events in the Northern Soul scene explicitly ban re-pressings from being played, and also that, in the hip-hop DJ tradition, 7″ records attract greater cache than the standard 12″ format. As such, we may observe the conspicuous nature of record collecting as the eventscape is designed to facilitate this swapping of subcultural capital (Thornton, 1995); SE noted that this formed a crucial part of the experience before and after performing, where participants enact an informal 'show and tell' ritual of comparing and contrasting their collections and taste. This, however, does not extend to the use of the decks themselves. All three interviewees offer help to performers unfamiliar with the decks (which, as double turntable systems, are different from a standard home set-up of one turntable); this is different from other music performance event where it is assumed that the performer will have the requisite competence to perform without assistance. This suggests that it is in fact the music which is the primary attraction and symbol of cultural capital, rather than the performance *per se*.

Community elements: Beyond this implicit system of kudos, however, the lack of musical 'policy' means that the potential community for each event is relatively large; indeed, all three interviewees explained the welcoming culture which they endeavour to implement at their respective events. All interviewees noted the development of (real/implicit) community feeling within their events not only relating to records but also on a personal support level (JF and MH). SE noted this interaction particularly in a cross-generational context, whereby participants from older generations had commonality with younger performers (e.g. Yang, Li & Zhao, 2017). This is significant as it exposes an interesting by-product of the re-emergence of vinyl whereby those who experienced vinyl as the only means of purchasing music meet those who consume vinyl as a retro, luxury or discretionary product. This appears to be a valuable element of the eventscape as it ensures that the event retains a contemporary relevance to participants in any age bracket (Yang, Li & Zhao, 2017). Indeed, a recurring sense of social tradition was evident, through the record as an enduring social artefact going back 70+ years (SE) (see Plasketes, 1992).

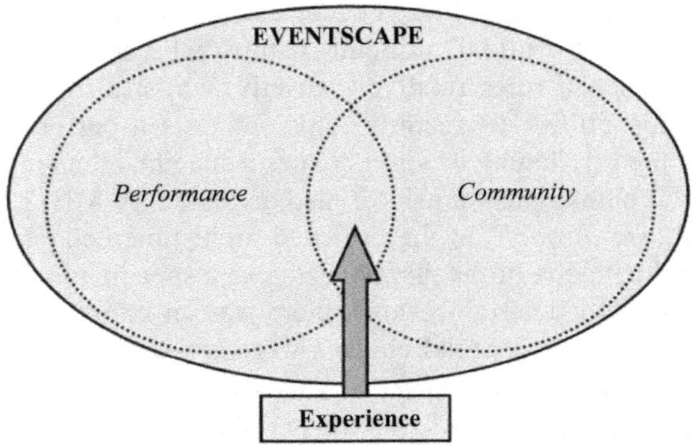

Figure 9.1 How are open decks events performed?

Finally, there is also the necessary social element of the open decks event; performers can listen to the music at home but choose to perform them in an environment which is by nature public. This suggests the desire to connect and contribute to the community; indeed, JF related this in an extreme case where an individual paid £90 for a record which he already owned in a different format:

> "You already had it?" "Yeah but I wanted to bring it down and play it!" (JF)

This positions the eventscape as an environment which is valued and important to the community in question, a concept echoed by MH in the way in which the event has been perpetuated by various volunteers and venue owners during its itinerant lifetime at different venues.

These two themes appear to constitute the key elements of experience and thus denote how the event is 'performed' both from a managerial and participatory perspective. This relationship is shown in Figure 9.1.

What well-being outcomes are achieved through open decks events?

Open decks events are not specifically a vehicle for well-being outcomes; these are arguably by-products of the central experience performed as explained previously. Nonetheless, two key well-being outcomes have been observed through this research in line with previous research in this field: community outcomes and personal well-being outcomes.

Community well-being outcomes: The role of community was a recurring theme within the research. Whilst 'community' can be a rather nebulous concept (e.g. Carter & Heim, 2015), within this research, it chiefly encompassed three elements: safety, commonality and conviviality, which therefore constitute the well-being outcomes from these events. These elements arose through multiple strands of interview discussion; indeed, all interviewees noted the place of their event within a wider open decks 'scene' in their respective locales. SE noted that many performers at his event travel a relatively long distance to contribute as that event forms part of an open decks/record collecting network. JF noted that his event coincides with another local open decks event, and many performers wish to perform at both in one night. MH also explained previous attempts to map the open decks scene in London through the Meetup web service not only to boost the visibility of his own event but also to stimulate a multi-event community and thus increase this crucial part of the event experience.

Through this sense of community, JF and MH both recounted instances where the event community has provided moral support for performers experiencing personal difficulties. However, the notion of a 'safe' environment extends beyond this into a sense of freedom to play records which would either be unwelcome or unheard in other environments. The free and 'safe' nature was specifically noted by MH and SE with MH seeking to create

> a safe space for anyone who liked very strange music (MH)

For SE, this manifested in a sense of convivial sharing, describing the environment as more '*intimate*' and '*personal*'; for him, this environment facilitated not only telling of 'stories' from person to person based around the records but also relating to life stories entwined with the record, its music and time (memory is key to this as discussed subsequently). Furthermore, all three interviewees saw the potential to meet new people and form new communities as a direct personal benefit for themselves and a motivation to promote their own event.

Whilst the scope for community is clearly strong, there is the danger that this attribute can become blurred or diminished; JF lamented that for a time his event became '*a bit too cool for school*' and at risk of losing its intended welcoming environment. This phase of the event passed in time; as such, the role of the host in facilitating and directing the community environment is an important element in ensuring these outcomes can be achieved. All interviewees noted the importance of welcoming performers, particularly new performers, and thus encouraging the community. This is made easier due to the relatively low barriers to entry for performers (contrasting, e.g. with open mic nights as noted previously), though MH noted that an open decks event which he attended before starting his own had a '*gatekeeper*' with the power to end any music deemed unsuitable. As such, the sense of community and the associated benefits, though widely identified in events (Sixsmith & Boneham, 2007; Vougioukalou *et al.*, 2019), cannot be taken as a necessarily organic outcome; they require ongoing facilitation by community members.

Personal well-being outcomes: This research identified multiple opportunities for feelings of well-being through personal attributes at open decks events. This can be seen through the prism of music experience which, as discussed

> in the literature review is noted as a stimulus for well-being. This is salient within the context of open decks events, specifically the concept of music performance (in a practical sense, rather than a 'performativity' sense).

Performance of records at the researched open decks events constitutes secondary performance (Frith, 2007). In this way, the focus is on the medium more than the performer, though by the nature of the events the performers' musical choices will be implicitly/explicitly judged and discussed by the audience. In this way, MH noted a link between open decks and open mic events, whilst JF described it as '*vinyl karaoke*', suggesting that the conventions and outcomes of those more established event types – performance, selection of music, cultural capital and quality – are present in open decks events. As such, the sense of 'performance' gives a sense of expression common in the playing of music (whether primary or secondary) (Frith, 2007), which has significant well-being benefits. SE described the arena as inherently one of '*self-expression*' through discussion and performance of records, giving the potential for eudaemonic feelings (Lamont, 2011), with the intimacy described by SE suggesting an element of risk and reward inherent in any performance (Frith, 2007). It also gives the performer a sense of active agency in contributing to the production of the event and, by extension, contribution to the life of the event community. Indeed, the sense of self-expression through sharing music with others and reciprocal performance has much in common not only with the established open mic format (Behr, 2012) but also more broadly the conventions of performing music in groups publicly or privately (e.g. Vougioukalou *et al.*, 2019); this suggests that performance at open decks events can have similar impacts on subjective well-being. Furthermore, JF noted the progression of performers in this respect; over time he has observed a personal growth in individuals from initially reluctant or unaccustomed to performance into confident regular performers. This sense of self-development through secondary performance is previously overlooked in events research and appears critical to understanding the potential benefits of open decks events as performance.

As such, we may see that the well-being outcomes come not only from the sense of performance but also from the community created in the eventscape. The relationship between the community and the individual also appears to reflect and create the sense of identity which all three interviewees explained as part of their own self-identity. This relatively recent event format gives an arena in which identity can be created, reflected and perpetuated through the prism of vinyl as the key signifier of open decks; therefore, the research now turns to the role of the vinyl records to better understand how these outcomes are affected by the central focus of open decks events.

What role do vinyl records play in the well-being outcomes?

To complete the conceptual framework of the study, this section will discuss the direct/mediating role which the record plays as a physical artefact and philosophical concept in the necessarily public domain of the open decks eventscape. It is noted that the record plays a physical role as well as an emotional role, combining into the significant yet intangible and elusive well-being stimulus herein termed the 'lure of vinyl'.

> *Physical role:* Regularly noted by the interviewees, the physicality of vinyl is its prime attribute as a social artefact. Part of this is the perception of vinyl as 'beautiful' (JF, MW, SE), something worthwhile transcending its musical contents when considering artwork (JF, MW, SE) (see Figure 9.2) and the cultural cache of records in 'immaculate' (JF) condition. The concept of 'object permanence' could explain this relationship, whereby the owner retains a conscious physical representation of music or the memory which it represents, enhancing its value not merely as a medium for playing music but also stimulating feelings of attachment, ownership (as opposed to streaming music which does not require ownership) and even a sort of 'relationship' with a record collection (MH). As such, we may regard vinyl as a 'stimulus' for the well-being outcomes, rather than as a direct cause. The physicality gives a focus for activity; chief amongst this, as evinced by all interviewees, is the concept of memory associated with vinyl records.

Rather than being memories themselves *per se*, vinyl records can be described as the embodiment of memory. The physicality becomes a vital 'touchstone of

Figure 9.2 'A lovely visual look': a Super Sevens performance with projection

Source: Reproduced with permission: Michael Hodgson.

memory' (Morgan & Pritchard, 2005); this could be related directly to the music itself (noted particularly by SE), but more generally there appears to be a preponderance of multidimensional memory which transcends the music and conveys social and personal memories through the prism of the record. For example, JF summarised this idea in a poignant quote which he linked to an image of himself with a friend performing at the open decks event:

> When you remember what you were, rather than what you've become (JF),

which points towards a deeper consideration of personal epochs (MH referred to 'childhood' memories), rather than simply considering when or where a song was first heard, or a record was purchased (though this is also an important aspect of the memory embodied by vinyl).

Emotional role: Whilst the physicality of vinyl is its chief attribute in stimulating well-being outcomes, it can also be shown to create emotional responses too specifically through the notion of 'attachment' to collections. MH explained that for many performers

> there's a lot of love for their record collections, the vinyl brings them a lot of happiness. (MH)

SE described records as being '*close to the heart'* of certain performers, and JF also expressed the potential for records to bring feelings of security or even salvation in the face of difficulty, suggesting a potential for a deep emotional response. Interestingly, however, JF reflected that this contemporary attachment to records contrasts with previous eras in which records were less valued or valuable, suggesting a more recent re-valorisation of the record in a well-being context. As happiness was not directly given as specific outcome, however, perhaps a more apposite emotional response to attribute to vinyl is eudaemonia. As noted in the literature review, the feeling of eudaemonia is a specific element of music consumption and one which was also noted by the interviewees in the context of open decks. This is different to happiness, suggesting a calmer sense of satisfaction rather than euphoria (Lamont, 2011) (though the two are not mutually exclusive). This is reflected most clearly in the sense of identity which some of the interviewees discussed; JF noted that vinyl has become a distinctive facet of his identity which is both created and reflected in his role managing an open decks event; SE also noted the record collection as an '*extension of the personality*', further explaining the role in identity.

The attachment also has a social memory function in repeating and maintaining social customs relating to the performance of music in this way. JF, as stated previously, noted the tradition of school discos where the collective records of children would be played; there was also repeated reference to nostalgia primarily from older performers (SE) experienced through the performance, ownership and discussion of vinyl. As such, the concept of memory appears to be central to several of the emotional outcomes from open decks – but also links to the presence of community as discussed previously. By facilitating this social history, the community within the eventscape, therefore, perpetuates the traditions and memories as part of a developing community, thus building the role of both the individual and collective emotional response.

> *'Lure of Vinyl':* This outcome represents a culmination of the preceding themes and sub-themes and embodies the often-elusive nature of vinyl as a social artefact in the 21st century, distinct from its pre-digital heyday as a more functional means of consuming music (Plasketes, 1992). The term 'lure of vinyl' was used by JF early in his interview and acted as an apposite 'anchoring' term for the subsequent research. JF described this is as something that '*can't be replicated*', is implicitly subsumed into memory, and which is '*important, memorable, resonant*'. This is grounded in the vinyl's physical manifestation, though that 'lure' mostly consists of intangible qualities. SE notes vinyl has an '*importance*' attached to it separate from the music contained on the disc; this extends to the distinctive technology (SE) which is considered interesting in itself despite its relative inconvenience (playback of vinyl causes degradation of the disc and also requires 'flipping' to play both sides) (MF, SE). Indeed, the physical ageing, despite the cache of '*immaculate*' records (JF) is seen as an endearing and important part of using vinyl (JF, SE) (Hayes, 2006). The physical qualities and the intangible qualities combine to create a complex polysemic artefact giving the user key agency in the use of the record; playing it releases not only the music but also the memories and, by extension, the well-being outcomes as discussed. It is this agency in utilising the object which differs from consuming music played by someone else, or even from other (i.e. digital) formats. Indeed, testament to this complexity is the relative difficulties which the interviewees had ineffectively summing up the essence of this 'lure'. By introducing this into the eventscape, the personal feelings are transplanted into a communal setting and thus contribute to the community outcomes as described.

As noted previously, this lure of vinyl can be seen as a postmodern *reification*; whereas in the pre-digital age music was necessarily consumed through a physical object (i.e. a record), the re-emergence of vinyl, and the display of this culture through open decks, is a retrospective process whereby record lovers ensure this physicality is not lost in an age where the vast majority of music is consumed in a digital and abstract manner – where the previously '*ordinary becomes special*' (JF). Ultimately, this positions the vinyl as not only a conduit for discrete well-being outcomes, as discussed herein, but also an unarticulated and elusive quality that gives this enduring medium its 'lure', as shown in Figure 9.3.

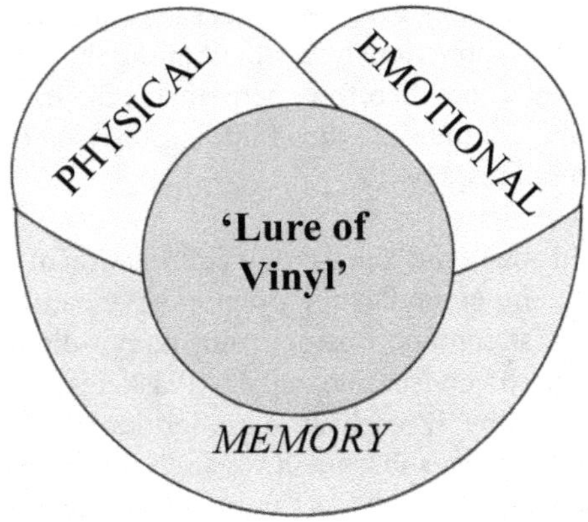

Figure 9.3 What role do vinyl records play in the well-being outcomes?

Conclusion

This research has introduced open decks events to the events studies literature to explore the potential for well-being outcomes. In doing so, it offers a fresh perspective on the potential for performance rituals to create and reflect meaning for the individual and collective. Through a focus on the vinyl record as a social artefact and active stimulus within this eventscape, the research examined the semiotic representation of well-being in the guise of what is termed herein the 'lure of vinyl'. This complex and intangible quality of the tangible object generates meaning through which we may establish well-being outcomes centred around the individual as a participant and performer within an implicitly/explicitly defined community.

The research, firstly, conceptualised this apparently contemporary event type and discussed the processes by which they are performed. Secondly, the types of well-being outcomes generated in the three examples were discussed through the personal and community dichotomy, and furthermore foregrounded the role of the vinyl record to understand how this enduring icon of music can stimulate and generate well-being for the individual and the community. By challenging the current academic conception of musical performance and participation and thus noting the agency involved in the performance of records, the outcomes of this research serve to offer more understanding in the burgeoning field of events and well-being research. Indeed, it is hoped that the findings will lead to further research in the role of the vinyl record as a paradoxically postmodern yet retrospective artefact of performance, memory and well-being. Figure 9.4 captures this in a moment where these elements appear to collide at Super Sevens.

Music can be timeless and, as evinced by countless musicians, it can make you smile and relieve the pain. Its *reification* as a 21st-century artefact in the

Figure 9.4 The lure of vinyl

Source: Reproduced with permission: Michael Hodgson.

contemporary eventscape suggests that, almost a century after its invention, the lure of vinyl may well be timeless too.

References

Armbrecht, J., & Andersson, T.D. (2019). The event experience, hedonic and eudaimonic satisfaction and subjective well-being among sport event participants. *Journal of Policy Research in Tourism, Leisure and Events*. Advance online publication. https://doi.org/10.1080/19407963.2019.1695346.

Bartmanski, D., & Woodward, I. (2015). The vinyl: The analogue medium in the age of digital reproduction. *Journal of Consumer Culture*. 15 (1), 3–27.

Bates, E. (2019). Vinyl as event: Record store day and the value-vibrant matter nexus. *Journal of Cultural Economy*, 1–19.

Behr, A. (2012). The real "crossroads" of live music – the conventions of performance at open mic nights in Edinburgh. *Social Semiotics*. 22 (5), 559–573.

Bennett, A., & Rogers, I. (2016). Popular music and materiality: Memorabilia and memory traces. *Popular Music and Society*. 39 (10), 28–42.

Boer, D., & Fischer, R. (2010). Towards a holistic model of functions of music listening across cultures: A culturally decentred qualitative approach. *Psychology of Music*. 40 (2), 179–200.

Bosnjak, M., Brown, C.A., Lee, D.J., Yu, G.B., & Sirgy, M.J. (2016). Self-Expressiveness in sport tourism: Determinants and consequences. *Journal of Travel Research*. 55 (1), 125–134.

Carter, D., & Heim, C. (2015). Community engagement or community conversation? "Boomtown", a large-scale regional, outdoor community theatrical event. *Australasian Drama Studies*. 66, 202–224.

Clift, H., & Hancox, G. (2010). The significance of choral singing for sustaining psychological well-being: Findings from a survey of choristers in England, Australia and Germany. *Music and Health*. 3 (1), 79–96.

Deery, M., & Jago, L. (2010). Social impacts of events and the role of anti-social behaviour. *International Journal of Event and Festival Management*. 1 (1), 8–28.

Devine, A., Quinn, B., & Devine, F. (2020). Reaching across the divide – the role of cultural events in peacebuilding. *Event Management*. DOI: 10.3727/152599519X15506259856552.

Frith, S. (2007). Live music matters. *Scottish Music Review*. 1 (1), 1–17.

Gabrielsson, A. (2002). Emotion perceived and emotion felt: Same or different? *Music Scientae*. 5 (1), 123–147.

Getz, D., & Page, S.J. (2019). *Event studies: Theory, research and policy for planned events* (4th ed.). London: Routledge.

Grebenar, A. (2020). Lost in music: Mapping the 21st century house music event experience. *Journal of Policy Research in Tourism, Leisure and Events*, 1–19.

Hacking, S., Secker, J., Kent, L., Shenton, J., & Spandler, H. (2006). Mental health and arts participation: The state of the art in England. *Journal of The Royal Society for the Promotion of Health*. 126 (3), 121–127.

Haley, D., & Senior, P. (2007). Art, health and well-being. In: Haworth, J., & Hart, G. (eds.) *Well-being: Individual, community and social perspectives*. Basingstoke: Palgrave Macmillan, pp. 110–126.

Haslam, D. (2016). *Life after dark: A history of British nightclubs and venues*. London: Simon & Schuster.

Hayes, D. (2006). "Take those old records off the shelf": Youth and music consumption in the postmodern age. *Popular Music and Society*. 29 (1), 51–68.

Haynes, G. (2017, July 27). The reissue racket: How many more 'classic' albums will be repackaged? *The Guardian*. Available from www.theguardian.com/music/2017/jul/28/reissue-racket-how-many-classic-albums-repackaged [Accessed 30 June 2020].

Helfet, G. (2019, January 4). 4.2 million records sold in the UK during 2018. *The Vinyl Factory*. Available from https://thevinylfactory.com/news/4-2-million-records-sold-in-the-uk-during-2018 [Accessed 30 June 2020].

Hemmett, D. (1998). Dangerous dancing and disco riots: The northern warehouse parties. In: McKay, G. (ed.) *DiY culture: Party & protest in Nineties Britain*. London: Verso.

Henry, J. (2007). Positive psychology and the development of well-being. In: Haworth, J., & Hart, G. (eds.) *Well-being: Individual, community and social perspectives*. Basingstoke: Palgrave Macmillan, pp. 25–40.

Hesmondhalgh, D. (2000). The British dance music industry: A case study of independent cultural production. *British Journal of Sociology*. 49 (2), 234–251.

Jepson, A., Stadler, R., & Spencer, N. (2019). Making positive family memories together and improving quality-of-life through thick sociality and bonding at local community festivals and events. *Tourism Management*. 75, 34–50.

Kawase, S., & Obata, S. (2016). Psychological responses to recorded music as predictors of intentions to attend concerts: Emotions, liking, performance evaluations, and monetary value. *Musicae Scientiae*. 20 (2), 163–172.

Lamont, A. (2011). University students' strong experiences of music: Pleasure, engagement, and meaning, *Musicae Scientiae*. 15 (2), 229–249.

McDermott, O., Orrell, M., & Ridder, H.M. (2014). The importance of music for people with dementia: The perspectives of people with dementia, family carers, staff and music therapists. *Aging and Mental Health*. 18 (6), 706–716.

Morgan, N., & Pritchard, A. (2005). On souvenirs and metonymy. *Tourist Studies*. 5 (1), 29–53.

Morrissey. (1987). 'Paint a vulgar picture'. Available on: 'Strangeways, here we come', by The Smiths. Warner Music UK. (Track Eight).

O'Grady, A. (2015). Dancing outdoors: DiY ethics and democratized practices of well-being on the UK alternative festival circuit. *Dancecult*. 7 (1), 76–96.

Packer, J., & Ballantyne, J. (2011). The impact of music festival attendance on young people's psychological and social well-being. *Psychology of Music*. 39 (2), 164–181.

Palm, M., (2017). Analog backlog: Pressing records during the vinyl revival. *Journal of Popular Music Studies*. 29 (4), 1–14.

Petterson, R., & Getz, D. (2009). Event experiences in time and space: A study of visitors to the 2007 world Alpine Ski championships in Åre, Sweden. *Scandinavian Journal of Hospitality and Tourism*. 9 (2–3), 308–326.

Plasketes, G. (1992). Romancing the record: The vinyl de-evolution and subcultural evolution. *Journal of Popular Culture*. 26 (1), 109–122.

Richards, G., de Brito, M.P., & Wilks, L. (2013). *Exploring the social impacts of events*. London: Routledge.

Rodgers, N., & Edwards, B. (1979). 'Everybody dance'. Available on: 'Chic' by Chic. Atlantic. (Track Four).

Sixsmith, J., & Boneham, M. (2007). Health, well-being and social capital. In: Haworth, J., & Hart, G. (eds.) *Well-being: Individual, community and social perspectives*. Basingstoke: Palgrave Macmillan, pp. 75–92.

Stebbins, R.A. (2018). Leisure and the positive psychological states. *Journal of Positive Psychology*. 13 (1), 8–17.

Thornton, S. (1995). *Club cultures: Music, media and subcultural capital*. Cambridge: Polity Press.

van der Hoevens, A., & Hitters, E. (2019). The social and cultural values of live music: Sustaining urban live music ecologies. *Cities*. 90, 263–271.

Vougioukalou, S., Dow, R., Bradshaw, L., & Pallant, T. (2019). Well-being and Integration through community music: The role of improvisation in a music group of refugees, asylum seekers and local community members. *Contemporary Music Review*. 38 (5), 533–548.

Weinberg, M.K., & Joseph, D. (2017). If you're happy and you know it: Music engagement and subjective well-being. *Psychology of Music*. 45 (2), 257–267.

Wood, E.H., & Moss, J. (2015). Capturing emotions: Experience sampling at live music events. *Arts and the Market*. 5 (1), 45–72.

Yang, A., Li, E., & Zhao, M. (2017). Harvard crooners: Building intergenerational relationships through participatory music making. *Journal of Intergenerational Relationships*. 15 (4), 419–422.

Yolal, M., Gursoy, D., Uysal, M., Kim, H. (Lina), & Karacaoğlu, S. (2016). Impacts of festivals and events on residents' well-being. *Annals of Tourism Research*. 61, 1–18.

Zadeh, J. (2014, October 28). You're getting totally ripped off by reissues, box sets and deluxe editions. *VICE*. Available from www.vice.com/en_uk/article/ryg8nm/why-most-reissues-are-rip-offs [Accessed 30 June 2020].

10 Music festivals

An immersive experience. How are emotions, well-being and cultural identity interrelated in event experiences?

Jordi Oliva and Alba Colombo

Introduction

Over the last few decades, social interest in individual and group well-being has risen. As a result, any elements, processes and actions related to the analysis or improvement of well-being have also garnered great interest. In this regard, the analysis of cultural events as potentially positive experiences linked to well-being has been studied from different perspectives, for example, in relation to community and its impacts (Phipps & Slater, 2010; Yolal *et al.*, 2016), to the participation of young people (Packer & Ballantyne, 2011) and to the benefit for family well-being (Jepson & Stadler, 2017). As a type of cultural event, music festivals have triumphed economically and socially in recent decades. These events constitute a growing business for the music sector. This is the case in Spain, for instance, where the number of festivals has almost doubled in the last decade. Specifically, in 2005, there were 551 festivals, while, in 2018, there were 896 (APM, 2005–2019).

Scholars focusing on event studies have often linked festival experiences in terms of expectations, motivations, consumer behaviour and commercial interests, among other factors. However, we strongly believe that attendees' experiences are highly related to emotions, which are currently influenced by the festival profile, the festival context or the main culture exposed. Attendee's experiences are not simply individual acts, but rather group processes that may be related to personal or social well-being. In addition, different events (sporting events, commercial events, cultural or traditional events, etc.) all have their own unique characteristics; cultural festivals also differ according to the cultural sector they target and their type, genre and style. Moreover, within each event type, there can be multiple differentiations depending on artistic style or added cultural tendency. For example, music festivals can be extremely diverse: imagine the differences between an open-air festival, a series of classical concerts across several theatres, and an experimental electronic music festival in an abandoned warehouse. Therefore, the diversity of these events may lead to a multiplicity of experiences, based on different elements such as atmosphere, social interaction and location, even if they all share a musical common ground (Packer & Ballantyne, 2011). It is also popularly recognised that certain types of music festivals may be associated with negative effects or experiences related to drugs (Jenkinson *et al.*, 2014; Martinus *et al.*, 2010) or violence (Fileborn, Wadds & Tomsen, 2020). Whatever the case may be, we believe that impacts should be identified according to the characteristics of the event in question due to the influence of these elements in the conformation of the experience of the festivalgoer.

There is a lack of research observing the relationship between event experiences (considering different event characteristics), emotions and well-being. Moreover, it could also be considered that the uniqueness of each experience is related to specific strong musical experiences. Thus, certain moments of intensity related to music (Gabrielsson, 2002) may help to shape the experience into an engaging activity. This engagement, and the resulting intense positive emotions felt in the context of the cultural activity, may provide a special moment in time that improves the attendee's individual well-being. In this chapter, our main argument is that participation in music festivals may generate the kind of experience that gives rise to positive emotions, engagement and accomplishment, all of which lead to enhanced personal well-being.

Literature review

Event experience can be understood under the umbrella of the 'experience economy' (Bell, 1973; Berridge, 2007; Jackson, 2006; Pine & Gilmore, 1999; Toffler, 1980). Within the experience economy, several analyses have suggested that event providers offer unique and remarkable experiences (Manthiou *et al.*, 2014; Mehmetoglu & Engen, 2011; Oh, Fiore & Jeoung, 2007; Park, Oh & Park, 2010). Specifically, several authors consider that music festivals shape unique experiences for each individual attendee (Ballantyne, Ballantyne & Packer, 2014). This experience has been analysed through different perspectives and models. Two such examples are de Geus, Richards and Toepoel (2016), who proposed the Event Experience Scale (EES), and Delamere, Wankel and Hinch (2001) and Small, Edwards and Sheridan (2005), who developed a theory for understanding the assessment of event's social impact. However, these two models centred their attention on impacts, singling out impact intensity as the most important feature of an experience, without identifying what really intensified the experience or made it unique. De Geus *et al.* (2016) used their EES to describe experience as a mixture of four components: affective, cognitive, physical and the experience of novelty. The EES operationalisation concluded that 'affective engagement' (the excitement, emotional energy, intimacy, sense of adventure, values and recollections attendees gain from an event, as stated in de Geus *et al.*, 2016) is one key factor. Other authors have underlined emotions related to satisfaction, motivation and expectation, as well as how the experience might be interpreted (Wood & Moss, 2015). In light of the foregoing, emotions may be considered as essential in the study of experiences at live music events such as music festivals.

Emotions and music

Emotions can be described by their characteristics, which are identified by their origin or the circumstances in which they arise. Most relevant to our research is the notion of aesthetic emotions, which defines emotions as a key factor in processing and enjoying works of art (Menninghaus *et al.*, 2019). This definition implies the ability to describe emotions through a discrete list thereof. MacGregor and Müllensiefen (2019) verified the consistency of use of Ekman's (1992) basic emotions in music, and the ability of music to convey, evoke or regulate these. However, the classical, discrete point of view, which considered that the entire set of emotions can be derived from a basic set of emotions (Ekman, 1992, 1999), has proven insufficient in the case of music-elicited emotions (Scherer & Zentner, 2001; Zentner, Grandjean & Scherer, 2008). Aesthetic emotions, in this respect, refer to the additional emotions evoked through the ability of music to complement and enrich the wide range of art-elicited emotions. Along these lines, Zentner, Grandjean and Scherer (2008) carried out analyses in four quantitative and qualitative studies, the findings of which identified a discrete list of emotions called the Geneva Emotional Music Scales (GEMS). The GEMS proposed three scales of music-elicited emotions felt by listeners. The first scale comprised 40 terms referring to emotions. Through factor analysis, these were labelled as one of the nine overarching emotions: wonder, transcendence, tenderness, nostalgia, peacefulness, power, joyful activation, tension and sadness. This list was subsequently validated by Trost *et al.* (2012), who mapped these reactional emotions in different regions of the brain. Emotions related to musical experiences can thus be labelled following consistent previous analyses.

Every human emotion is linked to a cognitive process that introduces semantic content (Koelsch *et al.*, 2004); that is, new opportunities to add cultural content to one's cultural identity. However, this process may also be the result of the cultural construction of emotions. Mesquita, Boiger and De Leersnyder (2016) took this construction as the basis for several research endeavours, describing culture as the framework within which people collectively feel emotions. Understanding the link between emotion and culture thus becomes a key requisite for investigating the construction of cultural identity through music festival experiences, and the key question is, therefore, how to measure the effects or the significance of culture in music festival experiences.

Cultural impacts, peak experiences with music and well-being

Colombo (2016) proposed a new model (Cultural Impact Perception) for evaluating and classifying cultural impacts in the context of an event, basing her analyses on attendees' perceptions of impact and identifying a set of impacts and their relationship to organisers' intentions to this effect. Colombo's model specifies and defines cultural impacts on the basis of previous definitions and models from a number of other authors (Delamere, Wankel & Hinch, 2001; Delamere, 2001; Fredline, Jago & Deery, 2003; Getz, 1991, 2007; Hall, 1989; Small, Edwards &

Sheridan, 2005). Colombo's (2016) classification introduces the possibility of identifying cultural impacts on different manifestations of culture in the context of cultural events.

Positive social and cultural impacts within the context of cultural events can be evaluated as opportunities for enhancing attendees' well-being. Yolal *et al.* (2016) observed how residents' participation in community festivals influenced their subjective well-being. In their analysis, they incorporated items related to socio-cultural benefits and costs, as proposed by Delamere, Wankel and Hinch (2001) and considered by Colombo (2016). These items were related to similar elements used in Diener, Horwitz and Emmons (1985) to refer to positive emotions or feelings. Their results showed a positive relationship between community benefits and residents' subjective well-being. This was the first methodological attempt to measure feelings-related items in relation to items concerning the sociocultural benefits afforded to attendees. However, the analysis conducted by Yolal *et al.* (2016) was limited by the method used to specify emotions evoked by the festival experience. Moreover, the design of unique experiences aims to generate specific moments that will be recalled as special in the future. These special circumstances or special moments in life were described as peak experiences by Maslow (1954), who would go on to correlate these peak experiences with the uppermost level of his hierarchy of needs: 'self-transcendence', through which the individual transcends their own personal concerns and views things from a higher perspective, moving beyond 'self-actualisation' or self-fulfilment and the realisation of one's potential. Maslow suggested that intense life experiences could also function as peak experiences capable of facilitating positive growth. This perspective, whereby fulfilment is achieved through highly engaging activities, was proposed in Csikszentmihalyi's (1990) definition of the 'flow' state of optimal experience. Csikszentmihalyi's investigations showed similarities with Maslow's theories in that special physical, cognitive and transcendental moments are expressed as peak experiences capable of contributing to the realisation of human potential. This understanding is comprehended in the holistic scope of the humanistic psychology (Aanstoos, Serlin & Greening, 2000) in the study of human behaviour offering an expanded and rigorous horizon of methods of inquiry and the conceptualisation of the human condition. In this scope, a broad, systematic analysis with music was not provided until the descriptive work of Gabrielsson (2001, 2002, 2011), which took Maslow (1968), Panzarella (1980) and Laski (1961) into consideration but adopted an open methodological approach and was not restricted to classical music. Whereas Maslow and Panzarella focused on positive or joyous experiences, Gabrielsson and Lindström (2003, p. 163) was interested in 'any kind of strong experiences which may occur in connection with music', which the author referred to as 'Strong Musical Experiences' or SME. Subsequently, Lamont (2011, 2012) was able to associate SME with an increase in well-being and the potential to make positive memories.

Later on, Schäfer, Smukalla and Oelker (2013, p. 529) – on the basis of the Strong Music Experience proposed by Gabrielsson – defined an intense musical experience (IME) as 'an unforgettable experience with music that stands out from usual experiences because of its quality; it is characterised by changed perception, deep feelings, and an intense physical reaction'. According to this definition, these moments are important because of their deeper, individual significance and the fact that they may be interpreted as a description of an engaging activity. The existence of these intense musical experiences in the context of the music festival may therefore be understood as an engaging activity for the festivalgoer.

In addition, the importance of participating in such engaging activities was demonstrated as a way to enhance the subjective well-being analysed by Seligman in his book on positive psychology *Flourish: A Visionary New Understanding of Happiness and Well-being* (2011). Seligman proposed five essential elements necessary for the enhancement of well-being: positive emotions, engagement, relationships, meaning and accomplishment (PERMA). Experiencing positive emotions may be essential for well-being (Seligman, 2011; Webster, 2014), and these are elements which may be felt during the music festival experience. The second element of PERMA, engagement, refers to the powerful link that may exist between an individual and an activity. This state can be defined in terms of the flow experience of playing an instrument (Csikszentmihalyi, 1990) or the benefits of learning and playing (Custodero, 2002; Lee, Krause & Davidson, 2017). The third element, relationships, concerns the capacity or ability of an individual to build positive relationships with others (Seligman, 2011). 'Meaning' refers to the individual's understanding that their life has purpose, and the final element, 'accomplishment', is associated with 'meaning' in that a purposeful life will lead to a sense or feeling of accomplishment. This understanding from the positive psychology is the framework of the analysis developed in the following sections.

Methodology

The aim of this research is to relate the emotional outcome of music festival experiences with the positive enhancement of well-being through three elements of the PERMA framework: positive emotions, engagement and accomplishment (Seligman, 2011). Emotional outcome is measured with the GEMS scale (Zentner, Grandjean & Scherer, 2008) and the evaluation of intense musical experiences (IME) proposed by Schäfer, Smukalla and Oelker (2013). This tool gathered data about specific emotions and peak experiences related to positive emotions and engaging in intense activities within the festival experience.

The sense of accomplishment is then deduced from the identity development evaluated by the impact on cultural identity construction, measured with the model proposed by Colombo (2016), which evaluates the existence of five cultural impacts through the perceptions of festivalgoers. Table 10.1 presents descriptive concepts of each impact; this content will be contrasted with qualitative data gathered from interviews. Each impact has a positive and negative consideration drawn from the analysis of the literature on the social and cultural impacts performed by Colombo (2016). Analysis of these cultural impacts may help to determine the specific impact on cultural identity construction and hence

Table 10.1 Cultural impacts and their descriptive concepts

Benefits		*Costs*	
Impact	*Items*	*Impact*	*Items*
Information about culture	Exposure to a variety of cultural experiences through the community festival Shared experience	Disinformation about culture	Negative community image
Preservation of cultural traditions	Revitalisation of traditions Preservation of traditions	Loss of cultural traditions	Loss of language Loss of heritage Alteration of traditions
Construction of cultural identity	Validation of community groups Impact on regional cultural identity Building of community pride Opportunity to develop new cultural skills and talents Celebration of community Impact on local community character Increased local interest in regional culture and history	Loss of cultural identity	Cultural profanation Loss of cultural amenities
Integration through cultural effects	Community pride and integration Cultural integration	Creation of ghettos as a result of cultural effects	Cultural marginalisation Community alienation
Social cohesion through culture	Opportunity for intercultural contact Community groups working together to achieve common goals through the festival Variety of cultural experiences	Social exclusion through culture	Cultural offence Loss of community pride and division Social dislocation Potential for intercultural misunderstanding

Source: Colombo (2016).

understand whether this can generate a sense of accomplishment in terms of self-cultural development.

With this three-fold evaluation, we were able to understand the music festival experience in relation to the promotion of the subjective well-being of the festivalgoers.

The first of the three PERMA elements was analysed as an experience of the positive emotions (P) according to the positive emotions contained in the GEMS-9 list of emotions, composing the emotional effect (Zentner, Grandjean & Scherer, 2008). The evaluation of these emotions in the context of the music festival also considered whether the emotions of the GEMS-9 appeared to be perceived as positive or negative.

Seligman's (2011) second element, engagement, refers to intense or engaging activities in which the individual experiences complete absorption and full self-involvement. These experiences may be understood from different perspectives, but since our context of analysis is music-related activity, our approach is based on an actual model called intense musical experience (IME), as discussed previously.

The third element, the sense of accomplishment described by Seligman (2011) as a way to improve subjective well-being, is strongly linked to cultural identification. The festival experience can be described as an opportunity to develop festivalgoers' cultural identity, as an individual accomplishment. Hence, identity development or sense of self-accomplishment by cultural means would demonstrate a sense of accomplishment, a qualitative increase in the festivalgoer's well-being.

To be able to apply these models of analysis, we adopted a mixed-methods approach, using both qualitative and quantitative data, to enrich and ensure the data were fit-for-purpose. To do so, between 2017 and 2019, we first developed a survey containing specific questions on the existence of the types of experience to be evaluated. We distributed the survey to attendees of three music festivals, who participated in the study as volunteers. Festival sizes varied from medium (approximately 2,000 participants) to large (approximately 120,000 participants). Quota sampling was performed, obtaining a total of 148 answers, from 68 women and 80 men. The festivals were carefully selected so that, even though they all involved musical performance and were all held in the Basque Country and Catalonia, the festival profiles were very different: Bilbao BBK Live, an open-air rock and pop music festival; Estival de Jazz, an experimental, community-based music festival held in Igualada (Barcelona); and Jazzaldia, a jazz festival held in San Sebastian. The survey was analysed following the parameters of the previously presented models summarising the emotional effect, which is the intensity according to a Likert scale of each of the nine emotions of the GEMS-9, and the existence of intense musical experience through a definition of the concept and nine specific categories of IME drawn from Schäfer, Smukalla and Oelker (2013). Table 10.2 presents the final constitution of the main terms of evaluation of the emotional effect.

We then interviewed ten festivalgoers from the same region as that of the three festival case study locations who had attended a number of different festivals during the past year, in Table 10.3 are presented. The classification of the festival type in this case was performed according to the musical genres referred to in Schäfer and Sedlmeier's (2009) classification of musical preferences.

These interviews were coded and clustered into main themes. Focusing on the cultural impacts, a deductive content analysis, which structured the codes in

Table 10.2 Operationalisation of the emotional effect

Dependent Variables	*Independent Variables*	*Indicator*
Emotion	Wonder Transcendence Power Tenderness Nostalgia Peacefulness Joyful activation Sadness Tension	1 to 5
IME	Existence of IME (with the definition) Oceanic experience Visionary restructuring Anxiety-free dissolution of the I Absence of thoughts Mental reflection Ineffability Deep inner experience Affective dominance Sensitisation Emotional learning	Yes/No

Source: Own elaboration, based on Schäfer, Smukalla and Oelker (2013) and Zentner, Grandjean and Scherer (2008).

Table 10.3 Interviewee profiles

Reference	*Age*	*Gender*	*Festival Type*	*Pseudonym*	*Origin*
R0304	20	M	Electronic[1]	Ray	Aragon
F0405	23	M	Alternative	Fabián	Navarra
J0104	25	M	Alternative	Jay	Cantabria
P0104	26	M	Alternative	Paul	Castile–Leon
M0503	30	W	Alternative	Marta	Catalonia
R0503	31	M	Alternative	Roger	Catalonia
T2004	31	W	Alternative	Tina	Catalonia
A0409	37	W	Alternative	Audrey	Catalonia
A2804	45	W	Sophisticated	Adele	Basque Country
R1704	57	W	Sophisticated	Rose	Valencia

Source: Own elaboration.

relation to the definitions of the cultural impacts, was performed. A total of 130 different statements were compiled under the category of cultural impact.

Results

The application of mixed data gathering methodologies in the different cases generated heterogeneous data that required separate analysis. First and foremost, the quantitative analysis requires a conceptual clarification between positive and negative emotions, which will be addressed in the next section. Following this clarification, the description of the variables regarding emotions is interpreted and related to Seligman's concepts of well-being, the importance of positive emotions to the experience of the festival, and the existence of intense musical experience as a manifestation of taking part in engaging activities. Finally, in the qualitative perspective, the results of the interviews show how a sense of achievement related to cultural identity development was obtained, thanks to the specific statements of the festivalgoers describing their experiences in interviews.

Positive emotions and engagement in the music festival

Seligman's (2011) understanding of the positive well-being brought about by an individual's perception of positive emotions is also presented by Diener, Horwitz and Emmons (1985), who relate the frequency and duration of positive feelings to the enhancement of subjective well-being. It is therefore essential to identify the types of emotion related to the specific experience of the festival and be able to determine whether these contribute to the improvement of subjective well-being. As previously stated, this selection was performed in relation to experiences at three music festivals, according to a Likert scale, for each emotion identified on the GEMS-9. As introduced previously, the GEMS-9 offers a scale composed of the following emotions: wonder, transcendence, tenderness, nostalgia, peacefulness, joy, power, tension and sadness (Zentner, Grandjean & Scherer, 2008).

The differentiation between positive and negative emotions should be addressed within the terms of the GEMS-9. Intuitively, the emotions of wonder, transcendence, tenderness, peacefulness, joy and power were considered positive, whereas tension and sadness were regarded as negative. On the one hand, the differentiation of positive and negative affect has been demonstrated in precedent studies (Watson, Clark & Tellegen, 1988; Watson & Naragon, 2009), which validated the difference and existence of positive and negative affective responses with empirical data. On the other hand, the terms 'tension' and 'sadness' have been included in other emotions scales: the differential emotions scale or DES (Izard *et al.*, 1993); the positive and negative affect schedule or PANAS-X (Watson & Clark, 1999); and the profile of mood states or POMS (McNair, Lorr & Droppleman, 1971), which have been confirmed through respective analysis and argumentation. By evoking primarily positive emotions, therefore, the festival experience may contribute to improved well-being on the part of attendees.

Existence of positive emotions

In our empirical study, having applied a Likert scale (from 1 to 5, where 1 is the least intense and 5 is the most intense), the results showed higher scores for the positive emotions in all three of the festivals analysed. The most intense emotions were joy (3.85 points), wonder (3.69) and power (3.55), depending on the type of the festival, and the least intense were tension (2.07) and sadness (1.53).

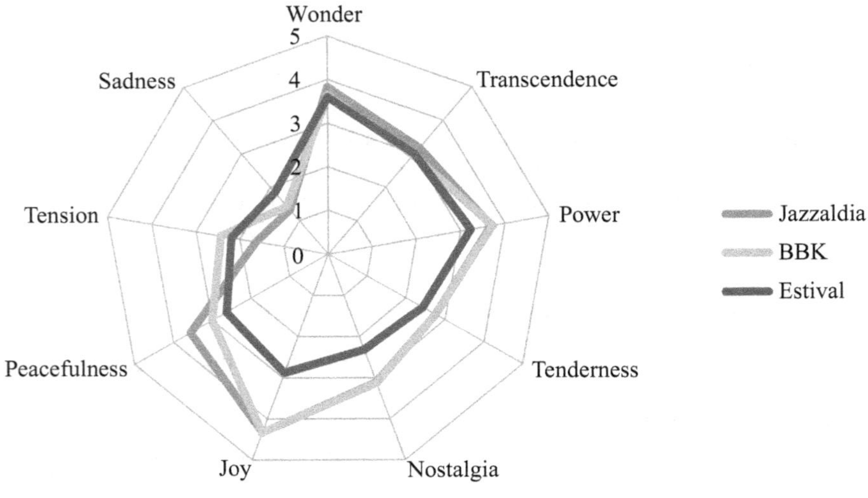

Figure 10.1 Distribution of means of emotions in the three festivals analysed

Analysis of this data may help us to generate a map of the emotions by applying the GEMS-9 within the festival experience.

As it is shown in Figure 10.1, the most intense emotions were recorded for Jazzaldia and BBK Live in relation to joy (4.33 and 4.34 points, respectively), while the least intense at the same two festivals was sadness (1.42 and 1.33, respectively). At Estival de Jazz, the most intense emotion was wonder (3.59), while the least intense was also sadness (1.82). Statistical coherence cannot be maintained in a comparison since the sample is not statistically representative. It is possible, however, to identify a trend in the importance of positive emotions such as joy and wonder over negative emotions such as sadness.

Consequently, it can be interpreted that positive emotions, evaluated using GEMS-9, were perceived with greater intensity by attendees of these music festivals. The music festival experience, as an experience involving positive emotions, can therefore be formulated as a potential way to promote a positive construct of well-being. Nevertheless, further research will be required to investigate the implications of the low significance of the positive emotion of tenderness (with a value of 2.65) and the ambivalent emotion of nostalgia (with a mean value of 2.85), which can be interpreted as either positive or negative (Barrett *et al.*, 2010).

Existence of peak experiences

As introduced previously, another essential element of the positive consideration of Seligman's (2011) PERMA theory is engagement. Seligman maintained that the 'engaged life' is an important element in the development of an individual's sense of positive well-being. In this section, these types of experiences are

Table 10.4 Descriptions of responses according to a festival

Festival	*Existence of IME*		*Non-existence of IME*	
	N	%	*N*	%
BBK Live	21	42%	29	58%
Jazzaldia	37	69%	17	31%
Estival de Jazz	33	77%	10	23%

Source: Own elaboration.

analysed using the survey, including a number of specific definitions proposed by Schäfer, Smukalla and Oelker's (2013, p. 529) intense musical experiences (IME). The results are heterogeneous for the three festivals analysed.

As given in Table 10.4, in the case of Jazzaldia and Estival de Jazz, existence of IME is reported by a higher number of respondents (69% and 77%, respectively), whereas in the case of BBK Live, the results are different, with respondents divided as to the existence or non-existence of IME (42% and 58%, respectively). This suggests that differences between IME at the different festivals may be related to the festival profiles and that the BBK Live experience may be less engaging than the two jazz-type festivals.

In relation to festival typologies, it is worth mentioning that some of the responses regarding the BBK Live festival may be linked to the higher rating of the GEMS-9 emotion category of tension. This festival has several characteristics that make it different from others, including aspects that may influence attendees' emotions, such as overcrowding, type of music and queuing, all of which are factors that may have the effect of blocking intense or peak experiences during this type of open-air camping festival. Nevertheless, our case studies show that the music festival experience provides opportunities for the eliciting of positive emotions and, in some cases, intense or peak experiences through music. Therefore, we consider that music festival experiences may generate IME and give rise to engagement as part of an 'engaged life'.

Cultural identity development and accomplishment

As presented previously, the qualitative approach focuses on evaluation of the cultural impacts generated by festivals, as this may give us a perspective which will enhance our understanding of the positive effects of festivals on participants. Following a conceptual explanation of the different cultural impacts and their manifestations, interviewees were asked to select and elaborate on their perceptions of cultural impacts during the music festivals. This deductive process helps us to identify the most relevant cultural impacts as well as understand the nuances of the experience from the perspective of the festivalgoer. On the one hand, we observe the quantification of statements by interviewees in relation to each code and, on the other hand, we identify relevant statements.

As stated in Table 10.2, the quantitative relevance of each category of 'Cultural impact' is shown in the number of statements and the number of different interviewees who made statements on each specific topic.

As represented visually in Table 10.5, the most quantitatively relevant impact is the one on the cultural identity construction, rated by 49% of the interviewees. In addition to being the most selected, the interviewees refer to this impact throughout the interview. Different codes then referred to specific themes that were extensively related to this impact. These themes also allow us to comprehend the festival experience as an opportunity to develop cultural skills, musical interests and talents, as well as build a sense of community and construct cultural identity.

The following qualitative analysis shows us that some statements are clearly related to quality of life and hence may help us to understand their relevance in terms of well-being enhancement. For example, it is considered that the music festival experience is an opportunity to develop cultural skills, musical interests and talents. The music festival is a place of cultural development whereby music and social experience contributes to a specific type of identity construction, as indicated by Fabián:

> The event contributed culturally. First of all, because culture is constructed by people. You can see people, you meet an assortment of people during the event, that contributes.
>
> (F-0405, 30:16)

Others expressed that the festival experience is an opportunity to take part in something culturally relevant, as indicated by Jay:

> I think it provides something cultural because people open their minds there.
>
> (J-0104, 29:36)

In the same vein, these statements may also express that the experience, during the festival, may be part of a larger process which transforms attitudes, processes

Table 10.5 Number of statements per interviewee and percentage of total statements

Impact	*Total Statements*	*Number of Interviewees*	*Percentage*
Construction of cultural identity/negative construction of cultural identity	64	10	49%
Cohesion/exclusion	40	10	31%
Information/misinformation	28	10	21.5%
Preservation/loss of traditions	10	8	8%

Source: Own elaboration.

and beliefs for some attendees. Thus, the festival experience may contribute to increased interest; for instance, the music festival is an opportunity to increase interest in unknown musical styles or bands, as expressed by Jay and Marta:

> You can discover and you may have a few days (after the festival) where you are listening to the artists you have seen, because you have discovered a song by a group that you liked and want to listen to it again.
>
> (J-0104, 15:00)

> After the festival, you tend to listen to more music and have greater musical appreciation.
>
> (M-0503, 17:10)

As music festivals are understood as a unique opportunity for cultural enrichment, festivals may provide opportunities for individuals to identify themselves with a particular musical style. Music festival experiences may thus contribute to the process of cultural identification, which is in turn part of the process of self-growth and self-development through culture.

Some of the interviewees' statements are related to the idea of building community, togetherness and even a sense of belonging. These ideas emerged from descriptions that emphasised the significance of the social experience, connection with others, and pride in being a member of the festival audience. As described by Jay, the festival itself is a kind of community:

> Yeah, you feel like you like the same music those people like. So you do feel part of a group, in a sense. It's like going to watch football: if you go to watch a game, you know those people like to watch football. So if you go to a certain festival, you like that music just like the others, then you're there as a community.
>
> (J-0104, 26:37)

Hence, as expressed in this statement, a music festival can be an occasion to share cultural values and engendering feelings of unity with others perhaps not considered culturally close. These elements may contribute to a process of building community pride which is linked to the city or region in which the festival is organised, as expressed by Paul:

> Yeah, since I started going to festivals. My friends don't like the same music as me. But I tell them we're going to a festival! They'd rather go to Ibiza for a week than go to a festival. Now, they're starting to go to festivals.
>
> (P-0104, 17:17)

Finally, interviewees also report identity development, and this is justified by two general observations: the cultural learning processes involved in music festival experiences and the cultural or personal shifts produced within this festival experience. For example, some interviewees considered the music festival an opportunity for cultural learning as well as to learn how to be socially involved. As Audrey reports:

> I believe that it always contributes something to the culture, . . . you learn many things: how to live in society, see the values, how to live as a person and find a voice.
>
> (A-0409, 19:01)

Music festivals also provide an arena for change evoked by musical experience, as illustrated in Adele's statement below. Ray goes further, expressing the idea that the intense experiences lived within the music festivals helped to construct identity:

> It changed my life completely, it was such a strong impact for me that it made me want to dedicate myself to singing and work in the performing arts industry.
>
> (A-2804, 12:08)

> The experiences we have define us, and the stronger the experience, the more it defines you, that's how it is with a festival.
>
> (R-0304, 39:23)

Music festivals can therefore provide an opportunity to have a set of profound learning experiences and contribute to self-development, from a small shift to an important contribution to an individual's self-understanding.

The interviewees' statements describe the festival experience as an opportunity to learn, develop cultural skills, increase individual interest in music and produce cultural and personal shifts. The festival experience can therefore be interpreted as a personal accomplishment in the construction of the individual's cultural identity, as suggested by Hand and Gresalfi (2015), who described collective accomplishment between individuals and their interactions with their cultural context. These were also related to a sense of accomplishment, which contributes to subjective well-being enhancement (Lee, Krause & Davidson, 2017; Seligman, 2011). Moreover, accomplishment can be understood as an opportunity to experience self-growth, as described by Lawendowski and Besta (2018). This sense of self-construction occurring during participation in a music festival may be understood as a sense of purpose or accomplishment. Consequently, the festival experience contains elements that enhance the construction of positive well-being through an activity related to the sense of accomplishment.

Conclusions

As the goal of this chapter is to relate the emotional outcome of a music festival experience to the subjective well-being characteristics described in Seligman's PERMA theory, the cases analysed indicate that such experiences provide opportunities for festivalgoers to feel positive emotions, fully engage in an activity and experience a sense of achievement. As mentioned previously, this can be explained by the fact that the levels of intensity described by attendees were higher for the positive emotions. Moreover, the existence of intense musical experiences in the context of the festival shows that the music festival experience contains special moments of intense engagement. Moreover, these positive emotions not only were common at different types of music festivals, as in the case studies analysed, but also were expressed heterogeneously by a sample of festivalgoers. Therefore, the music festival experience in the examples analysed contains fundamental elements of the subjective well-being construction described by Seligman (2011) and endorsed by Diener (2009), through the existence of intense positive emotions and intense engaging experiences with music, as we have shown.

Festivalgoers also revealed their experience of cultural development through the subjective description of perceived cultural impacts in the semi-structured interviews. Interviewees described identity development within music festival experience, development being understood as the opportunity to acquire a sense of achievement through the experience, this being linked to enhanced subjective well-being.

Although further research is needed, the results of our analysis may serve to break new ground in understanding emotions within the framework of music festival experience, and how these emotions relate to personal well-being. Our contribution is the appreciation that attending a festival, specifically a music festival, may be a positive personal experience, providing the opportunity to experience a number of positive emotions capable of contributing to personal or collective satisfaction and well-being.

Taking into consideration the limitations of this research, a deeper analysis will be required to observe the complex role played by a music festival-elicited experience in the personal self-identification or even self-construction of an individual. For example, what exactly happens at an open-air camping festival (e.g. BBK Live), which seems to have a less engaging effect on attendees? Similarly, we could investigate whether different music festival profiles influence attendees' emotions, not only in terms of music styles or genres but also taking into account other factors such as context, venue, atmosphere and the sociodemographic profile of festivalgoers. In sum, despite our endeavour to explore the link between music festival experience and well-being, further research is needed to evaluate and identify profoundly different kinds of positive emotions with different festival profiles to verify that relationship. Our exploration of the idea that participation in a music festival may contribute to enhanced well-being through elements such as positive emotions, engagement and sense of accomplishment has left us with the profound conviction that the music festival experience may contribute to life satisfaction and therefore subjective well-being.

Note

1 The classification of the festival type in this case was performed according to the musical genres referred to in Schäfer and Sedlmeier's (2009) classification of musical preferences.

References

Aanstoos, C.M., Serlin, I., & Greening, T. (2000). A history of Division 32 (humanistic psychology). In: Dewsbury, D.A. (ed.) *Unification through division: Histories of the divisions of the American Psychological Association*, Vol. V. Washington, DC: American Psychological Association, pp. 85–112.

Anuario de música en vivo. *Asociación de Promotores Musicales de España*, APM 2005–2019. (2020, April 10). Madrid. Spain. Available from www.apmusicales.com/category/apm/anuario/.

Ballantyne, J., Ballantyne, R., & Packer, J. (2014). Designing and managing music festival experiences to enhance attendees' psychological and social benefits. *Musicae Scientiae*. 18 (1), 65–83.

Barrett, F.S., Grimm, K.J., Robins, R.W., Wildschut, T., Sedikides, C., & Janata, P. (2010). Music-evoked nostalgia: Affect, memory, and personality. *Emotion*. 10 (3), 390–403.

Bell, D. (1973). *The coming of post-industrial society: A venture in social forecasting basic books*. New York: Heinemann Educational Publishers.

Berridge, G. (2007). *Events design and experience*. Abingdon: Routledge.

Clark, G. (2008). Local development benefits from staging global events. Paris: OECD Publications.

Colombo, A. (2016). How to evaluate cultural impacts of events? A model and methodology proposal. *Scandinavian Journal of Hospitality and Tourism*. 16 (4), 500–511.

Csikszentmihalyi, M. (1990). *Flow: The psychology of optimal experience*. New York: Harper & Row.

Custodero, L.A. (2002). Seeking challenge, finding skill: Flow experience and music education. *Arts Education Policy Review*. 103 (3), 3–9.

de Geus, S., Richards, G., & Toepoel, V. (2016). Conceptualisation and operationalisation of event and festival experiences: Creation of an event experience scale. *Scandinavian Journal of Hospitality and Tourism*. 16 (3), 274–296.

Delamere, T. (2001). Development of a scale to measure resident attitudes towards the social impact of community festivals: Part II: Verification of the scale. *Festival Management and Event Tourism*. 7, 25–38.

Delamere, T., Wankel, L.M., & Hinch, T.D. (2001). Measuring resident attitudes toward the social impacts of community festivals. Pretesting and purification of the measure. *Festival Management and Event Tourism*. 7, 11–24.

Diener, E. (2009). *Assessing wellbeing. The collected works of Ed Diener*. Berlin: Springer.

Diener, E., Horwitz, J., & Emmons, R.A. (1985). Happiness of the very wealthy. *Social Indicators Research*. 16, 263–274.

Ekman, A.K. (1999). The revival of cultural celebrations in regional Sweden: Aspects of tradition and transition. *Sociologia Ruralis*. 39 (3), 280–293.

Ekman, P. (1992). An argument for basic emotions. *Cognition and Emotion*. 6, 169–200.

Fileborn, B., Wadds, P., & Tomsen, S. (2020). Sexual harassment and violence at Australian music festivals: Reporting practices and experiences of festival attendees. *Australian & New Zealand Journal of Criminology*. 53 (2), 194–212.

Fredline, E., Jago, L., & Deery, M. (2003). The development of a generic scale to measure the social impacts of events. *Event Management*. 8, 23–37.

Gabrielsson, A. (2001). Emotions in strong experiences with music. In: Juslin, P.N., & Sloboda, J.A. (eds.) *Music and emotion: Theory and research*. New York: Oxford University Press, pp. 431–449.

Gabrielsson, A. (2002). Emotion perceived and emotion felt: Same and different. *Musicae Scientiae*. 10 (2), 123–147.

Gabrielsson, A. (2011). *Strong experiences with music: Music is much more than just music*. New York: Oxford University Press.

Gabrielsson, A., & Lindström, S. (2003). Strong experiences related to music: A descriptive system. *Musicae Scientiae*. 7 (2), 157–217.

Getz, D. (1991). *Festivals, special events and tourism*. New York: Van Nostrand Reinhold, 374 p.

Getz, D.(2007). *Event studies. Theory, research and policy for planned events*. Oxford, England: Butterworth-Henemann. Elsevier.

Hall, M. (1989). The definition and analysis of hallmark tourist events. *GeoJournal*. 19 (3), 263–268.

Hand, V., & Gresalfi, M. (2015). The joint accomplishment of identity. *Educational Psychologist*. 50 (3), 190–203.

Izard, C.E., Libero, D.Z., Putnam, P., & Haynes, O.M. (1993). Stability of emotion experiences and their relations to traits of personality. *Journal of Personality and Social Psychology*. 64 (5), 847–860.

Jackson, C. (2006). The experiential impact of events. In: Fleming, S., & Jordan, F. (eds.) *Events and festivals: Education, Impacts and Experiences*. Eastbourne: Leisure Studies Association Publications, pp. 131–145.

Jenkinson, R., Bowring, A., Dietze, P., Hellard, M., & Lim, M.S.C. (2014). Young risk takers: Alcohol, illicit drugs, and sexual practices among a sample of music festival attendees. *Journal of Sexually Transmitted Diseases*, 1–6.

Jepson, A., & Stadler, R. (2017). Conceptualizing the impact of festival and event attendance upon family quality of life (QOL). *Event Management*. 21 (1), 47–60.

Koelsch, S., Kasper, E., Sammler, D., Schulze, K., Gunter, T., & Friederici, A.D. (2004). Music, language and meaning: Brain signatures of semantic processing. *Nature Neuroscience*. 7 (3), 302–307.

Lamont, A. (2011). University students' strong experiences of music : Pleasure, engagement, and meaning. *Musicae Scientiae*. 15 (2), 229–249.

Lamont, A. (2012). Emotion, engagement and meaning in strong experiences of music performance. *Psychology of Music*. 40, 574–594.

Laski, M. (1961). *Ecstasy. A study of some secular and religious experience*. London, England: Cresset Press.

Lawendowski, R., & Besta, T. (2018). Is participation in music festivals a self-expansion opportunity? Identity, self-perception, and the importance of music's functions. *Musicae Scientiae*. https://doi.org/10.1177/1029864918792593.

Lee, J., Krause, A.E., & Davidson, J.W. (2017). The PERMA wellbeing model and music facilitation practice: Preliminary documentation for wellbeing through music provision in Australian schools. *Research Studies in Music Education*. 39 (1), 73–89.

Macgregor, C., & Müllensiefen, D. (2019). The musical emotion discrimination task: A new measure for assessing the ability to discriminate emotions in music. *Frontiers in Psychology*. 10 (AUG), 1–15.

Manthiou, A., Lee, S., Tang, L., & Chiang, L. (2014). The experience economy approach to festival marketing: Vivid memory and attendee loyalty. *Journal of Services Marketing*. 28 (1), 22–35.

Martinus, T., McAlaney, J., McLaughlin, L.J., & Smith, H. (2010). Outdoor music festivals: Cacophonous consumption or melodious moderation? *Drugs: Education, Prevention and Policy*. 17 (6), 795–807.

Maslow, A.H. (1954). *Motivation and personality*. New York: Harper & Row.

Maslow, A.H. (1968). *Toward a psychology of being* (2nd ed). Oxford, UK: D. Van Nostrand.

McNair, D.M., Lorr, M., & Droppleman, L.F. (1971). *Manual for the profile of mood states (POMS)*. San Diego, CA: Educational and Industrial Testing Service.

Mehmetoglu, M., & Engen, M. (2011). Pine and Gilmore's concept of experience economy and its dimensions: An empirical examination in tourism. *Journal of Quality Assurance in Hospitality & Tourism*. 12 (4), 237–255.

Menninghaus, W., Wagner, V., Wassiliwizky, E., Schindler, I., Hanich, J., Jacobsen, T., & Koelsch, S. (2019). What are aesthetic emotions? *Psychological Review*. 126 (2), 171–195.

Mesquita, B., Boiger, M., & De Leersnyder, J. (2016). The cultural construction of emotions. *Current Opinion in Psychology*. 8, 31–36.

Oh, H., Fiore, A.M., & Jeoung, M. (2007). Measuring experience economy concepts: Tourism applications. *Journal of Travel Research*. 46 (2), 119–132.

Packer, J., & Ballantyne, J. (2011). The impact of music festival attendance on young people's psychological and social wellbeing. *Psychology of Music*. 39, 164–181.

Panzarella, R. (1980). The phenomenology of aesthetic peak experiences. *Journal of Humanistic Psychology*. 20, 69–85.

Park, M., Oh, H., & Park, J. (2010). Measuring the experience economy of film festival participants. *International Journal of Tourism Sciences*. 10 (2), 35–54.

Phipps, P., & Slater, L. (2010). *Indigenous cultural festivals: Evaluating impact on community health and wellbeing*. Melbourne: Globalism Research Centre, RMIT University.

Pine, B.J., & Gilmore, J.H. (1999). *The experience economy: Work is theatre & every business a stage*. Boston: Harvard Business Press.

Schäfer, T., & Sedlmeier, P. (2009). From the functions of music to music preferences. *Psychology of Music*. 37, 279–299.

Schäfer, T., Smukalla, M., & Oelker, S. (2013). How music changes our lives: A qualitative study of the long-term effects of intense musical experiences. *Psychology of Music*. 42, 525–544.

Scherer, K.R., & Zentner, M. (2001). Emotion effects of music: Production rules. In: Juslin, P., & Sloboda, J. (eds.) *Music and emotion: Theory and research*. Oxford, England: Oxford University Press, pp. 361–392.

Seligman, M.E.P. (2011). *Flourish: A visionary new understanding of happiness and well-being*. New York: Free Press.

Small, K., Edwards, D., & Sheridan, L. (2005). A flexible framework for evaluating the socio-cultural impacts of a (small) festival. *International Journal of Event Management Research*. 1 (1), 66–77.

Toffler, A. (1980). *The third wave: The revolution that will change our lives*. London: Collins.

Trost, W., Ethofer, T., Zentner, M., & Vuilleumier, P. (2012). Mapping aesthetic musical emotions in the brain. *Cerebral Cortex*. 22 (12), 2769–2783.

Watson, D., & Clark, L.A. (1999). The PANAS-X: Manual for the positive and negative affect schedule-expanded form. *University of Iowa. Psychology Publications*. https://doi.org/10.17077/48vt-m4t2.

Watson, D., Clark, L.A., & Tellegen, A. (1988). Development and validation of brief measures of positive and negative affect: The PANAS scales. *Journal of Personality and Social Psychology*. 54, 1063–1070.

Watson, D., & Naragon, K. (2009). Positive affectivity: The disposition to experience positive emotional states. In: Lopez, S.J., & Snyder, C.R. (eds.) *Oxford library of psychology. Oxford handbook of positive psychology*. New York: Oxford University Press, pp. 207–215.

Webster, A.D. (2014). A flourishing future: Positive psychology and its lessons for education. *Independent School*. 74, 40–46.

Wood, E.H., & Moss, J. (2015). Capturing emotions: Experience sampling at live music events. *Arts and the Market*. 5 (1), 45–72.

Yolal, M., Gursoy, D., Uysal, M., Kim, H. (Lina), & Karacaolu, S. (2016). Impacts of festivals and events on residents wellbeing. *Annals of Tourism Research*. 61, 1–18.

Zentner, M., Grandjean, D., & Scherer, K.R. (2008). Emotions evoked by the sound of music: Characterization, classification, and measurement. *Emotion*. 8 (4), 494–521.

Part II

Events and countering negative well-being

11 Well-being through art

A critical reflection on how a body image art event can improve the health and well-being of gay men

Phillip Joy

Introduction

This chapter tells stories. My story and the stories of other men in my community. These are the stories of the struggles we have with our bodies that influence and shape our mental health and well-being. In this chapter, I use principles of post-structuralism and queer theory in critical reflection, to weave my story with the stories of the men that participated in an art event. In many ways, our stories are reflections. I begin with a brief overview of body image concerns for gay men and how such concerns are created through social and cultural constructs of heteronormativity, stigma, and body ideals. I also briefly review the consequences of body dissatisfaction on the mental and emotional health and well-being of gay men. Then, I explore the literature that suggests art events may be a way to improve health and well-being. Following the review of the literature, I provide context to the research study and the resulting art event. The final sections of the chapter provide some of my personal reflections on how the process of designing and creating a community art show event was a way to create connections with others and help my own sense of well-being with my body.

Literature review

Body image, body dissatisfaction, and gay men

Body image is defined as the way we view and think about our bodies. Negative body image can drastically influence the well-being of people (Gattario & Frisén, 2019). Although there are many aspects to body image, research for gay men has primarily focused on fat and muscularity as main aspects of the bodies that influence the way men feel about their bodies (Varangis *et al.*, 2012). The meanings of fat and muscularity are socially and culturally constructed and create bodies that are privileged and more acceptable than other bodies. From a Western perspective, social discourses position a lean, muscular, and fat-free body as the ideal body for gay men. Images, social media and dating apps, television and pornography, and other cultural references continually (re)shape the meanings associated with different bodies for gay men (Joy, 2020). Cultural messages give meanings to bodies, influence how we understand ourselves, and can create pressures to embody standards of beauty. For many men embodying such body standards can be an impossible task (Gee & Jackson, 2017). It is the difference between our perception of our bodies and our understanding of dominant body ideals that create body dissatisfaction, tensions, and negative mental health and well-being (Bordo, 1999).

The consequences of body dissatisfaction to well-being can be immeasurable. Wellness can be thought of as a way of life in which a person seeks to live more fully a state of positive health and well-being and that considers the body, mind, and spirit (Myers, Sweeney & Witmer, 2000). In this definition, wellness is influenced and shaped by many factors including gender, community involvement, and cultural identity (Myers, Sweeney & Witmer, 2000). Body dissatisfaction can touch many areas of wellness from the sense of identity, to eating practices to sex lives. Numerous studies have documented the relationships between gay men and eating disorders, dieting practices, vomiting, excessive exercise practices, and steroid use to shape their bodies (Feldman & Meyer, 2007; Morgan & Arcelus, 2009; Brennan *et al.*, 2011; Lacefield & Negy, 2012; Kousari-Rad & McLaren, 2013; Mor *et al.*, 2014; Frederick & Essayli, 2016; Goedel *et al.*, 2017; Griffiths *et al.*, 2018). But beyond the physical consequences of body dissatisfaction are the mental, emotional, and social consequences which are less explored in the literature.

Dominant social discourses about the meanings of fat often position fat as unhealthy and people with fat as lazy, irresponsible, and undesirable (Puhl & Heuer, 2010; Longhurst, 2016; Trainer, Wutich & Brewis, 2017). For gay men, being labelled as fat or overweight often creates experiences of invisibility, isolation, and exclusion (Foster-Gimbel & Engeln, 2016). In a study by Edmonds and Zieff (2015), gay men spoke about their experiences of rejection and isolation from other gay men. Rejection and isolation based on bodies did not conform to the dominant muscular and lean ideals within gay culture. Participants in the Edmonds and Zieff (2015) study described feelings of being an outsider within the gay community that often lead them to withdraw from social situations, community events, and dances. This may be particularly isolating for gay men who are already marginalised within a heteronormative society. Feraios (1998) also described isolation from other gay men based on his body. He believed that people never saw him as a desirable gay man because of his weight. He felt invisible as he walked through the gay districts of New York and this negatively influenced his overall well-being.

Community events for wellness

Lesbian, gay, bisexual, transgender, queer, and two-spirited (LGBTQ2S) communities are often thought of as a group of similar individuals. But LGBTQ2S individuals are diverse. There are many LGBTQ2S communities that include people from different cultures, ethnicities, genders, and socio-economic backgrounds. Despite the many different communities, the health and wellness of the people who identify within these communities are continuously being (re)shaped by heteronormative social structures that create stigma and marginalisation (Henry *et al.*, 2020). Previous research indicates that people who cannot openly express their sexuality may have fewer social supports, leading to greater depressive symptoms and negative well-being (Slimowicz, Siev & Brochu, 2020). Experiences of community connections and supports through events, such as advocacy, volunteerism, art, and pop-up events, are means by which individuals within LGBTQ2S communities may improve their health and well-being. For example, advocacy events, such as pride events, can bring people together to change policies, disrupt social norms, and express their identities (Lewis, Small & Mehmet, 2019). Working towards common goals through community events has been reported to help gay men feel a sense of purpose, overcame experiences of stigma, develop a sense of belonging, feel connected to each other, and improve their sense of well-being (Ramirez-Valles *et al.*, 2005; Lewis, Small & Mehmet, 2019). As another type of event, pop-ups can include cultural performances, dances, food events, and art exhibits. Pop-ups can provide a temporary space for LGBTQ2S people to express and celebrate their identities, resist dominant social norms, connect with others, and building a sense of community (Stillwagon & Ghaziani, 2019), which are all integral for positive well-being (Myers, Sweeney & Witmer, 2000; LeBeau & Jellison, 2009; Wong *et al.*, 2014; Morandini *et al.*, 2015).

Art for well-being

Art and the process of creating art can be transformative (Asakura *et al.*, 2019). Research has shown that the use of art within health programs has a number of benefits, including decreasing feelings of isolation, improving self-esteem, developing and broadening social networks, and improvements in well-being (Crone *et al.*, 2013). Art engages the human senses, emotions, and creativity (Finley, 2008). Art has been used in health programs as a form of therapy to elicit self-expression and healing, at both the individual and community levels (Bien, 2005). Art can challenge social constructs of identity and create new knowledge and ways of being through the expression of new perspectives (Schuhmann, 2014).

Within the LGBTQ2S communities, art has been used extensively in community programs to improve social connections and well-being (Hebert *et al.*, 2017). Researchers have also utilised art within various methodologies, such as photovoice, as a means of knowledge translation and social change (Wang *et al.*, 1998; Carlson, Engebretson & Chamberlain, 2006; Hergenrather *et al.*, 2009; Catalani & Minkler, 2010). Photovoice is an arts-based methodology that aims to tell the stories of participants in hopes to create social change that improves the health and well-being of communities (Wang & Burris, 1994). The process of taking photographs becomes a way for people to share their individual and collective experiences. This method has been used to explore such topics as suicide in gay, bisexual, and two-spirited men (Ferlatte *et al.*, 2019), the experiences of queer and trans youth (Holtby *et al.*, 2015), the experiences of trans men accessing healthcare (Hussey, 2006), and body image concerns of gay men (Joy, 2020).

Methodologies

The aim of this chapter is to highlight, through my personal critical reflection, the way a community art event can build connections, enhance a sense of community, and contribute to feelings of improved well-being for gay men who experience tensions relating to their body image. The following section provides an overview of the theoretical perspectives that informed this work.

Critical reflection through a queer poststructural lens

Critical reflection as a research method involves the researcher adopting a reflective stance to examine their position with the research (Fook, 2011; Morley, 2014). The intent of using critical reflection is to explore how researchers participate in the discourses that shape the findings. In this process, the researcher explores the influence of their background, assumptions, values, beliefs, and attitudes on the research process and findings (Fook, 2011). The process is both deconstructive and reflective (Fook, 2011). Critical reflection allows the researcher to understand more deeply the social discourses, cultural norms, and relations of power that shape and influence them and their research. This process also allows for the possibility of different ways of knowing and seeing new possibilities for change (Hickson, 2016). These goals of critical reflection are, therefore, well-aligned with poststructural and queer theory perspectives, which are also deconstructive in nature.

Poststructuralism is a theory of language and discourses that deconstructs how they shape knowledge, identities, and influence the way people come to know themselves and the world (Agger, 1991; Cheek, 1999; Fendler, 2010). Discourses are everything that is known about a given topic and shape what is known. Discourses produce identities in ways that are not always known to the person but are always set within historical contexts (Fendler, 2010). Poststructuralism explores the multiple and shifting meanings behind discourses through the analysis of cultural texts, including art and photography (Agger, 1991). The examination of discourses can allow societal assumptions to be challenged (Agger, 1991; Cheek, 1999).

Queer theory also views identities as social constructions and explores how discourses and power relations shape and influence them (Spargo, 1999). Queer theory examines the process of "othering" and aims to disrupt and to dismantle notions of "normal" or "abnormal" identities (Spargo, 1999; Nic Giolla Easpaig *et al.*, 2014). For me, the process of critical reflection in this research is also about dismantling what I believe to be "normal" and disrupting my assumptions, beliefs, and values in relation to body image. The following section positions myself as the researcher and provides insight into my story, my perspectives, and my views of my body before the art event.

Purpose of project and photovoice

As previously mentioned, this chapter explores my reflections on a community art event. Specifically, this event was a photography show created and held as part of my graduate research on body image for gay men. The purpose of my research was political in nature. I sought not only to understand how gay men navigated their personal struggles with body image, but I also sought to disrupt dominant discourses about ideal bodies within the gay community. I thought about how to achieve these goals in ways that would be captivating for myself and others. I had read about the political and transformative power of photovoice during my dietetic internship and as an amateur photographer myself, I thought there was no better way to do my graduate work than to use photography, effectively combining my academic world with my passion for photography. I was quite excited for the process.

I was not disappointed in the photovoice process. The community-based participatory process had five major steps, including (1) introductory workshop, (2) the photography phase in which participants took photographs on the research topic, (3) personal interviews, (4) an art show planning workshop, and (5) the art show event. Although this chapter is focused primarily on step 5, the art show event, all steps in the process are critical for the success of photovoice events. For myself, and the other participants, each step was an integral part of our journey to find our way to a deeper sense of well-being through the connections we made with each other.

The participants

The participants of the study who became co-organisers of the art show were recruited from Halifax, Nova Scotia, Canada, a small city on the east coast. It is a university city, with several major institutions that attract many students. Halifax has a thriving art scene with many local artists, galleries, and performing venues. Halifax also has a very small gay community and tends to be conservative towards diverse sexual orientations. Stigma around being gay is still present within the city and many gay people found it hard to connect with others. One participant noted how loneliness is often a part of gay life in Halifax and rural Nova Scotia.

Nine self-identifying gay men consented to partake in the project and to photograph their attitudes, beliefs, values, and practices about food and their bodies. The project details and consent processes were reviewed during the introductory workshop. Many of the group wanted more details about what to actually photograph, but I intentionally did not give them too much direction. I wanted their voice in the photographs, not mine. I wanted them to tell their stories and not what I thought their stories to be. I tried to achieve a delicate balance of sharing ideas while encouraging them to fully explore their own ideas in the workshop. In talking with some of the men after they took their photographs, this balancing act seemed to be successful. In the interviews, the men guided me narratively through their creative processes and the meanings of their photographs. I remember thinking after the interviews how lucky I was that my participants were so thoughtful in their process. I was overwhelmed with the emotional impact of their photographs.

Although all of us shared the experience of being gay and living in Halifax, we were also a diverse group of men, including several university students. Jonathan, aged 23, was in his fifth year of undergraduate studies, Mohammed, aged 25, was a health professional student studying dentistry, and Oliver, aged 28, returned to school after attending an art and design college to follow his passion for biology. Ryan, aged 25, had recently graduated from university with a Bachelor of Science in health. The group also consisted of Scott, aged 30, who worked as an engineer and Mark, aged 32, who was a registered nurse. Daniel, aged 55, was an active member within the local gay community and a long-standing member of many gay organisations. Ralph, aged 55, had recently returned to Nova Scotia to retire after a long career in the banking industry. The final member of the group was Scott, aged 65. He also recently retired from a 30-year career in the food industry for health reasons. He was diagnosed with Bulbar Onset ALS, a disease that affected his chewing, swallowing, and speech processes. During the time of the project, he was unable to verbally communicate but used writing to communicate. It should be noted that the informed consent process gave the participants the option to use their real names in both the art event and the research reports. I thought this was crucial in not erasing or losing their voices. The names used are therefore a mix of both real names and pseudonyms, as instructed by the participants themselves.

The participants each had individual motivations for agreeing to participate in the photovoice process but a common thread that united most of us was the desire to create awareness and stimulate discussions on the pressures gay men experience in relation to their body images. Many of the participants struggled with issues surrounding their bodies, fatness, masculinity norms, and the idea of achieving a body socially deemed to be the most attractive to other men. The chance to explore their attitudes, beliefs, and values about food and their bodies, as well as the chance for them to express themselves creatively through a reflective group event was appealing to many of them. The aspect of developing the event as a group was also a motivation for some. For example, Oliver thought his participation would help him connect with other gay men. For other participants, their involvement in the event was not only for personal connections but also for political. For example, Mark said his

> goal was to share his story as a way of owning it and not being ashamed of it . . . about putting it out there, saying this is me, this is my experience and if somebody can relate to that in some way or even be inspired by it a little bit, then that would be amazing.

For Mark, participating in the photovoice process and the art show event was a way to be known, a way to tell his story, and a way for him to connect and inspire other gay men who might be struggling with similar body image tensions. I shared many of the same motivations with Mark and, as I listened to their stories, I saw myself reflected. To more fully understand my reason for this project and wanting to organise such an event, the following section explores my story and my position as a researcher, a gay man, and as someone who has struggled with body image tensions. It speaks of my critical reflections prior to the event.

My journey to the event

My journey to this research and the art event was also a personal journey. My journey began with a reflection on my motivations for the research and my desire to create an art event for gay men to discuss the influence of body image on their well-being. I also saw this research as a means to understand myself, my own feelings, and to explore what has shaped my body tensions as a gay man.

I grew up in a very small rural Nova Scotia village where being labelled as different was not a good thing, and I was definitely labelled as different from the other boys. This became even more obvious to myself and everyone else as I entered junior high school. My non-conformity to strict gender rules, my interests in all things with unicorns, my natural shyness, and my growing awareness of my attraction to boys othered me from all my friends and the other students in the school. It felt as if a target was painted on my back. It was as if everyone smelled my otherness and it drew them to me like blood in the water. The constant day-to-day bullying lasted for the next three years of junior high school. Stress, anxiety, fear, and isolation were my companions. My mental, physical, and nutritional health suffered. Lunchtime for me was not about eating with friends, it was about escape and avoidance from my tormentors. I spent my lunchtime eating in the library where no one could find me. Eating became emotional for me. My body changed, becoming larger, rounder, and fatter. As the bullies tired of me, I was discovering the gay world. But even within that world, I felt isolated. The bodies of gay men that I saw in my magazines and on television were not my body. I was not tall, muscular, and had no rippling abs. My body was not deemed desirable by myself or by the other gay men I met. My struggles with my body intensified and still follow me to this day. I constantly compare my body to the bodies of other men both in the media and within the real world. I feel these struggles not only persist but also affect most aspects of my life, including my interactions with others, my relationships, and now my research.

Planning the art show event

Several weeks prior to the art show event, the participants and I met to discuss the planning of the event. This was a critical component of the event. It gave the men ownership of the event and how their photographs would be shown. It was very important to me that the men were invested in the show and led the planning of it. This was their art show event and I truly wanted them to be excited for the event. My philosophy was that if they saw themselves reflected not only in the art in the show but also in the show itself, they would enjoy the event more.

The planning workshop was attended by six of the participants. The beginning of the workshop was used to plan the overall details of the art show, including the date, time, location, and methods for promotion. I wanted to ensure everyone could attend and have plenty of time to invite friends and family. The rest of this planning workshop focused on how to present the participants' images to the audience. Budget was one consideration in the planning of the event. It was important for me that the participants did not incur any costs from the event. Going into the planning workshop I felt restrained by my limited budget. I believed I would not be able to print and display every photograph that the group submitted to me since altogether I had received about 85 photographs. Therefore, during the planning workshop, I asked the participants to look through their collective work and to select the one image that best expressed their story. I explained that for both logistical and financial reasons I thought printing their strongest photograph was the best way to design the event. I further clarified that I would print, matt, and frame this one photograph for the art show. They could create a title and a brief statement that would be displayed next to the image. This idea, however, was not well received by the group. They believed that only showing a single image from their series of photographs was not enough to tell their full stories. They told me that their stories were more complex than one photograph could capture. It was not a hard sell on their part as I fully understood their view. It was their event and the stories being told at the event were their stories. If my idea did not tell their stories effectively than what was the point? The idea of a single photograph from each of them for the event was rejected. At this point, I will admit that I was getting nervous on how to proceed, but the group was creative and determined to fully tell their stories. We brainstormed several ideas and it was collectively decided that each participant would create a collage with as many of their photographs as they wanted and that this collage would be printed as one large image that would be framed. I ran with this idea and thought it was brilliant. I asked each of the participants in the planning workshop to use small prints I had printed some long paper, markers, and tape to do a mock-up of their collage.

During their mock-up creation, I also suggested that we included some text in the show to provide some context to their photographs. The group thought this was a good idea and we discussed the best way to do this with the new collage idea. It was decided that a booklet would be created for attendees to read during the art show. The participants either wrote new statements for the booklet with their mock collage or gave me permission to use quotes from their interviews that explained the images. It was decided that each image within the collage would be given a reference number that corresponded to a number in the booklet with the text. We ended the planning workshop with a group discussion during which they shared their mock-ups collages and text. I felt it was important to give the participants time to talk and bond over their collective work. They left their mock-up collages with me to digitally re-create, print, and frame for the event. We also agreed that I would create the booklet using their templates. From my perspective, the planning workshop was integral to the participatory and advocacy nature of the art event. I believe the workshop allowed participants the autonomy to design their art pieces for the event, art pieces that would become the stimulus for creating discussions and disrupting hegemonic and oppressive body and gender norms for gay men.

The purpose of a photovoice art event is to bring awareness to the topic and to stimulate social disruption and change. I wanted to reach as many people as possible in the community, so I undertook promotion efforts that included listings in community newspapers, social media announcements, university notices, and two radio interviews on the Canadian Broadcasting Corporation. Local politicians and LGBTQ2S community activists were specifically invited and participants were asked to share the event details. I deeply wanted people to discuss the harmful effects body standards can have on people, particularly within the gay community.

Findings and discussion

Through the looking glass: the art show event

The art show, titled *Through the Looking Glass*, took place on January 10, 2019, from 6 pm to 9 pm in a local art gallery in Halifax, Nova Scotia. I invited all the participants to help me set up on the afternoon of the event and to join me for dinner as a way to celebrate the event. One participant, with his partner, helped me hang all the art pieces, move furniture, and set-up the music for the event. Another participant joined us for dinner. Looking back on this time reflectively, I was very glad that I extended this offer even though many of the participants were not able to help with the set-up or join us for supper due to prior engagements such as work and school. The opportunity to set up the gallery walls, to hang photos, to pick out music, to move furniture, and to share food with some of the participants was, I thought, a nice way for me to get to know these participants better – to bridge the gap between researcher and participant. When we were finished setting up the gallery, the space was transformed into a cozy, inviting, and welcoming space for people to talk about the artwork.

When the set-up was complete, we went to supper. Over food and drinks, we talked about the project and their thoughts about it. It was in these moments that I really, for the first time, got a sense of how nervous they were about the event and their photographs being on display for the whole world to see. I will come back to this thought later in my discussion as this became one of the main themes of my reflection on the event, the brevity of the participants in their expression.

With our bellies full, we went back to the gallery and opened the doors to the public. People started arriving. We greeted them with smiles and offered them the booklets to review as they moved through the artwork. It was a casual affair where people could drop by explore the artwork and stay for conversation, food, and connecting with the participating men. The event was well attended, with approximately 30–40 people attending, including executive board members of the local Pride Committee, local healthcare providers for gay men, and members of wider Halifax community (Figure 11.1). Throughout the event, several other of the participating men arrived with friends, family, and partners.

Many of the participants chose to identify themselves as the artists to the viewers, talking and sharing their thoughts on their artwork and their experiences, as seen in Figure 11.2. His story was of experiencing body image tensions – a story he was now able to share because of this involvement in this event.

Moments like the ones in Figure 11.2 made the event emotional for me. It was nice to see people connecting through their stories and their artwork. I felt that

Figure 11.1 Welcome table to the art show event, showing the guide booklet that provides quotes and stories for each piece of artwork

Source: Author.

Figure 11.2 Participant discussing his collage of photographs with an art show visitor

Source: Author.

this was a feeling that was shared by many of the people at the event. There were many hugs throughout the night and many attendees told me how powerful the stories and photographs were to them. A guest book was provided for visitors to leave comments (Figure 11.3). Feedback indicated visitors felt the photos, in combination with the booklet, were powerful, thought-provoking, and a meaningful way to translate knowledge about gay men's values, beliefs, and practices concerning food and bodies. The following are excerpts that illustrate the meaning of this event to people.

Figure 11.3 Sign in book for attendees of the art show event to leave messages

Source: Author.

> This was fantastic. I was especially moved by the stories that accompanied the photos.
>
> Visitor 1

> Beautiful work and so meaningful. It makes me hopeful, motivated and humbled to hear and see these perspectives, these stories.
>
> Visitor 2

> Such an amazing art show. Congratulation to all the participants. Photos and stories are beautiful and powerful.
>
> Visitor 3

> It was very interesting and thought-provoking. Keep encouraging this conversation.
>
> Visitor 4

> Beautiful body of work – thanks to all these brave men for sharing their life and journey. This is the best! Truly meaningful and intimate works.
>
> Visitor 5

The comments left by the visitors reveal that art events such as the one described here can be critical, deeply moving, and a thoughtful way to engage individuals, health leaders, and communities in creating conversations that challenge the social and cultural norms placed on the bodies of gay men. Other comments illustrate the way art events can also create new knowledge.

> This was great. It gave a great perception on things we would not think about each day. Thank you for opening our minds to this important topic.
>
> Visitor 6

> Thanks everyone for allowing us to see into and understand your experiences.
>
> Visitor 7

This event gave visitors the opportunity to think differently about bodies and how social expectations placed on bodies can impact the health and well-being of men. The event gave people the opportunity to expand their knowledge and understanding of the experiences of others.

This art event, in my reflection, was also beneficial to the participants in their journey of easing body tensions and finding a sense of well-being. In watching the participants engage with the visitors that night, in witnessing them talking about their struggles with body image with others, and in seeing them find understanding and compassion, was confirmation to me that the art event was a positive experience for them. One participant left a comment in the guest book. He said, *thank you for showing us as we were*. This comment illustrates how the event was appreciated by the participant and gave him the sense of being known. The event gave him the opportunity to shift his experiences of invisibility and non-representation within the gay community. It gave him the voice to say my body is fabulous and worth celebrating even if it is not represented or seen within dominant forms of media and within the dominant imagery of gay men's bodies. It took courage for him to show his authentic self in his photographs and it took vulnerability for him to show his photographs to complete strangers.

In my reflections of the event, these are the things that stand out – courage, authenticity, and vulnerability. All the men who took part in this project and captured intimate moments of their lives, who displayed their photographs and revealed their vulnerability and hurt to others during this event were courageous to me. They were brave to take part in this event, to talk about their experiences of hurt, shame, and isolation based on body standards. They were brave to talk about the anxiety, fear, and constant pressure they experience to embody the perfect body. They were brave to celebrate their bodies, whether their bodies conformed to, or challenged, dominant body standards for gay men. Authenticity can be defined as a comfort with one's identity and how honest a person can be to others with their identity (Riggle *et al.*, 2017; Petrocchi *et al.*, 2020). Although this definition, in reference to LGBTQ2S individuals, often refers to the acceptance of their sexual identity, I believe it can also be applied to one's comfort with one's body and one's identity that is created through socially constructed body ideals. Authenticity, self-awareness, and personal insights are beneficial to the well-being of individuals within the LGBTQ2S communities (Riggle & Rostosky, 2011) and I would argue that art events, such as this one, create opportunities for individuals to develop and work on all of these aspects of well-being. I would further argue that the participants in creating and participating in this event were engaging in an authenticity and that this speaks to the capacity of such events to contribute to an improved sense of well-being. The participants were authentic in their photographs. They revealed their bodies and said, through their artwork, this is who I am; this is what I struggle with; these are the things that have shaped me; these are the things I am accepting about myself.

The event also helped me with my body image and sense of well-being. It gave me perspective that other gay men share my struggles, that our stories are mirrored, and that we share so much more than just being gay. I found that I was not alone in my struggles with body image. I saw this reflected in the photographs from the other men. I heard this reflected in their interviews. I read this in their booklet quotes. I talked about this with them at the event. The realisation I was not alone was enough to help me in my own struggles. It was a deeper sense of community and belonging I felt with these men. The event helped me to feel less isolated in my experiences. A bond of common experiences that was also a bond of common purpose. The purpose to change social discourses about men's bodies so that others do not experience the negative consequences of dominant, rigid, and limiting body ideals for gay men. Increased sense of belonging (Riggle & Rostosky, 2011), increased social support from other gay men (Fingerhut, Peplau & Gable, 2010), and common advocacy goals (de Vries *et al.*, 2020) have all been suggested as contributing to improved well-being. These were the things I found with these other men and through participating in this art event. Participatory art events have been previously noted to help with reducing loneliness and social isolation within various populations, including LGBTQ2S peoples (Walters & Jepson, 2019). My new connections and sense of shared experiences with the other participants are the reason that my involvement in this art event contributed positively to my sense of well-being.

Conclusion

Social constructions of body ideals can negatively influence how gay men feel about their bodies and lower their overall sense of well-being. My personal experiences can attest to the influence of body image on well-being. But I am not alone in this. Many of the other men who took part in this project and its associated events reflected my personal experiences. There is a need to share these stories and, in doing so, find a way through our struggles to improved well-being. One way to do this is through art events. Participatory art events have been previously indicated to create community connections and relieve social isolation leading to improved feelings of well-being (Jepson, 2019). The *Through the Looking Glass* event provides more evidence on the benefits to well-being from the participation in art events. Through critical reflections, we are able to understand how art events can allow people to tell their stories and be known to others. Art events can be a means to create knowledge and new ways of understanding the experiences of others. Critical reflections through a queer poststructural theoretical lens provide us with a way to examine and explore social norms and discourses that shape our lives. Critical reflection can reveal new ways of knowing and different ways of living. The use of this critical reflection of the *Through the Looking Glass* event allows us to understand how participating in such events can create a sense of belonging, a connection with others that contributes positively to our well-being. I felt privileged to be a part of planning and implementing the *Through the Looking Glass* event with these other men.

Acknowledgements

I would like to acknowledge Dr. Laurene Rehman, Dr. Megan Aston, and Dr. Lisa Goldberg, Dr. Sara Kirk, and Dr. Matthew Numer, as well as financial support from the Canadian Institutes of Health Research (GSD-152289): the Frederick Banting and Charles Best Canada Graduate Scholarships Doctoral Award and the Izaak Walton Killam Pre-Doctoral Scholarships.

References

Agger, B. (1991). Critical theory, poststructuralism, postmodernism: Their sociological relevance. *Annual Review of Sociology*. 17 (1), 105–131.

Asakura, K., Lundy, J., Black, D., & Tierney, C. (2019). Art as a transformative practice: A participatory action research project with trans* youth. *Qualitative Social Work*. 19 (5–6), 1061–1077.

Bien, M.B. (2005). Art therapy as emotional and spiritual medicine for Native Americans living with HIV/AIDS. *Journal of Psychoactive Drugs*. 37 (3), 281–292.

Bordo, S. (1999). *The male body: A new look at men in public and in private*. New York: Farrar, Straus, and Giroux.

Brennan, D.J., Crath, R., Hart, T.A., Gadalla, T., & Gillis, L. (2011). Body dissatisfaction and disordered eating among men who have sex with men in Canada. *International Journal of Men's Health*. 10 (3), 253–268.

Carlson, E.D., Engebretson, J., & Chamberlain, R.M. (2006). Photovoice as a social process of critical consciousness. *Qualitative Health Research*. 16 (6), 836–852.

Catalani, C., & Minkler, M. (2010). Photovoice: A review of the literature in health and public health. *Health Education & Behavior*. 37 (3), 424–451.

Cheek, J. (1999). *Postmodern and poststructural approaches to nursing research*. Thousand Oaks, CA: Sage.

Crone, D.M., O'Connell, E.E., Tyson, P.J., Clark-Stone, F., Opher, S., & James, D.V. (2013). 'Art Lift' intervention to improve mental well-being: An observational study from UK general practice. *International Journal of Mental Health Nursing*. 22 (3), 279–286.

de Vries, J.M., Downes, C., Sharek, D., Doyle, L., Murphy, R., Begley, T., McCann, E., Sheerin, F., Smyth, S., & Higgins, A. (2020). An exploration of happiness within the Irish LGBTI community. *Journal of Gay & Lesbian Mental Health*. 24 (1), 40–76.

Edmonds, S.E., & Zieff, S.G. (2015). Bearing bodies: Physical activity, obesity stigma, and sexuality in the bear community. *Sociology of Sport Journal*. 32 (4), 415–435.

Feldman, M.B., & Meyer, I.H. (2007). Eating disorders in diverse lesbian, gay, and bisexual populations. *International Journal of Eating Disorders*. 40 (3), 218–226.

Fendler, L. (2010). *Michel Foucault*. Vol. 22. London: Continuum International Publishing Group.

Ferlatte, O., Oliffe, J.L., Salway, T., Broom, A., Bungay, V., & Rice, S. (2019). Using photovoice to understand suicidality among gay, bisexual, and two-spirit men. *Archives of Sexual Behavior*. 48 (5), 1529–1541.

Feraios, A.J. (1998). If only I were cute: Lookism and internalized homophobia in the gay male community. In Atkins, D. (ed.), *Looking queer: Body image and identity in lesbian, bisexual, gay, and transgender communities*. New York, NY: Harrington Park Press, pp. 415–429.

Fingerhut, A.W., Peplau, L.A., & Gable, S.L. (2010). Identity, minority stress and psychological well-being among gay men and lesbians. *Psychology & Sexuality*. 1 (2), 101–114.

Finley, S. (2008). Arts-based research. In: Knowles, J.G., & Cole, A.L. (eds.) *Handbook of the arts in qualitative research*. Thousand Oaks, CA: Sage, pp. 71–81.

Fook, J. (2011). Developing critical reflection as a research method. In: Higgs, J., Titchen, A., Horsfall, D., & Bridges, D. (eds.) *Creative spaces for qualitative researching*. Rotterdam: Sense Publishers, pp. 55–64.

Foster-Gimbel, O., & Engeln, R. (2016). Fat chance! Experiences and expectations of antifat bias in the gay male community. *Psychology of Sexual Orientation and Gender Diversity*. 3 (1), 63–70.

Frederick, D.A., & Essayli, J.H. (2016). Male body image: The roles of sexual orientation and body mass index across five national U.S. Studies. *Psychology of Men & Masculinity*. 17 (4), 336–351.

Gattario, K.H., & Frisén, A. (2019). From negative to positive body image: Men's and women's journeys from early adolescence to emerging adulthood. *Body Image*. 28, 53–65.

Gee, S., & Jackson, S. (2017). *Sport, promotional culture and the crisis of masculinity*. London: Palgrave Macmillan.

Goedel, W.C., Krebs, P., Greene, R.E., & Duncan, D.T. (2017). Associations between perceived weight status, body dissatisfaction, and self-objectification on sexual sensation seeking and sexual risk behaviors among men who have sex with men using Grindr. *Behavioral Medicine*. 43 (2), 142–150.

Griffiths, S., Murray, S.B., Krug, I., & McLean, S.A. (2018). The contribution of social media to body dissatisfaction, eating disorder symptoms, and anabolic steroid use among sexual minority men. *Cyberpsychology, Behavior, and Social Networking*. 21 (2), 149–156.

Hebert, P., Fernandez, R., Márquez, J., Ozne, E., Scott, A., La Cruz, D., Sixto, M., Vargas, A., & Warrant, D. (2017). Much better than a like: Creative pedagogies of wellness among young queer men of colour. *Visual Inquiry*. 6 (2), 209–229.

Henry, R.S., Perrin, P.B., Sawyer, A., & Pugh, M. (2020). Health conditions, access to care, mental health, and wellness behaviors in lesbian, gay, bisexual, and transgender adults. *International Journal of Chronic Diseases*, 1–8.

Hergenrather, K.C., Rhodes, S.D., Cowan, C.A., Bardhoshi, G., & Pula, S. (2009). Photovoice as community-based participatory research: A qualitative review. *American Journal of Health Behavior*. 33 (6), 686–698.

Hickson, H. (2016). Becoming a critical narrativist: Using critical reflection and narrative inquiry as research methodology. *Qualitative Social Work*. 15 (3), 380–391.

Holtby, A., Klein, K., Cook, K., & Travers, R. (2015). To be seen or not to be seen: Photovoice, queer and trans youth, and the dilemma of representation. *Action Research*. 13 (4), 317–335.

Hussey, W. (2006). Slivers of the journey: The use of photovoice and storytelling to examine female to male transsexuals' experience of health care access. *Journal of Homosexuality*. 51 (1), 129–158.

Jepson, A.S. (2019). 'De-marginalising' marginalised communities: The case of participatory arts events and the over 70s in rural Hertfrodshire, United Kingdom. In: Walters, T., & Jepson, A.S. (ed.) *Marginalisation and events*. London: Routledge, pp. 76–92.

Joy, P. (2020). *Through the looking glass: A poststructural and queer photovoice project exploring nutrition and bodies among gay men*. [Unpublished doctoral dissertation], Dalhousie University, Dalhousie.

Kousari-Rad, P., & McLaren, S. (2013). The relationships between sense of belonging to the gay community, body image dissatisfaction, and self-esteem among Australian gay men. *Journal of Homosexuality*. 60 (6), 927–943.

Lacefield, K., & Negy, C. (2012). Non-erotic cognitive distractions during sexual activity in sexual minority and heterosexual young adults. *Archives of Sexual Behavior*. 41 (2), 391–400.

LeBeau, R.T., & Jellison, W.A. (2009). Why get involved? Exploring gay and bisexual men's experience of the gay community. *Journal of Homosexuality*. 56 (1), 56–76.

Lewis, C., Small, F., & Mehmet, M. (2019). Together we will go our way: The development of a stakeholder framework for rural gay pride events. *Travel and Tourism Research Association: Advancing Tourism Research Globally*. 37, 1–12.

Longhurst, R. (2016). Queering body size and shape: Performativity, the closet, shame and orientation. In: Pausé, C., Murray, S., & Wykes, J. (eds.) *Queering fat embodiment*. Surrey: Ashgate Publishing, pp. 13–27.

Mor, Z., Parfionov, K., Davidovitch, N., & Grotto, I. (2014). Gym exercising patterns, lifestyle and high-risk sexual behaviour in men who have sex with men and in heterosexual men. *BMJ Open*. 4 (11), 1–8.

Morandini, J.S., Blaszczynski, A., Dar-Nimrod, I., & Ross, M.W. (2015). Minority stress and community connectedness among gay, lesbian and bisexual Australians:

A comparison of rural and metropolitan localities. *Australian and New Zealand Journal of Public Health*. 39 (3), 260–266.

Morgan, J.F., & Arcelus, J. (2009). Body image in gay and straight men: A qualitative study. *European Eating Disorders Review*. 17 (6), 435–443.

Morley, C. (2014). Using critical reflection to research possibilities for change. *British Journal of Social Work*. 44 (6), 1419–1435.

Myers, J.E., Sweeney, T.J., & Witmer, J.M. (2000). The wheel of wellness counseling for wellness: A holistic model for treatment planning. *Journal of Counseling & Development*. 78 (3), 251–266.

Nic Giolla Easpaig, B.R., Fryer, D.M., Linn, S.E., & Humphrey, R.H. (2014). A queer-theoretical approach to community health psychology. *Journal of Health Psychology*. 19 (1), 117–125.

Petrocchi, N., Pistella, J., Salvati, M., Carone, N., Laghi, F., & Baiocco, R. (2020). I embrace my LGB identity: Self-reassurance, social safeness, and the distinctive relevance of authenticity to well-being in Italian lesbians, gay men, and bisexual people. *Sexuality Research and Social Policy*. 17, 75–86.

Puhl, R.M., & Heuer, C.A. (2010). Obesity stigma: Important considerations for public health. *American Journal of Public Health*. 100 (6), 1019–1028.

Ramirez-Valles, J., Fergus, S., Reisen, C.A., Poppen, P.J., & Zea, M.C. (2005). Confronting stigma: Community involvement and psychological well-being among HIV-positive Latino gay men. *Hispanic Journal of Behavioral Sciences*. 27 (1), 101–119.

Riggle, E.D., & Rostosky, S.S. (2011). *A positive view of LGBTQ: Embracing identity and cultivating well-being*. Lanham: Rowman & Littlefield Publishers.

Riggle, E.D., Rostosky, S.S., Black, W.W., & Rosenkrantz, D.E. (2017). Outness, concealment, and authenticity: Associations with LGB individuals' psychological distress and well-being. *Psychology of Sexual Orientation and Gender Diversity*. 4 (1), 54–62.

Schuhmann, A. (2014). How to be political? Art activism, queer practices and temporary autonomous zones. *Agenda*. 28 (4), 94–107.

Slimowicz, J., Siev, J., & Brochu, P.M. (2020). Impact of status-based rejection sensitivity on depression and anxiety symptoms in gay men. *International Journal of Environmental Research and Public Health*. 17 (5), 1546–1556.

Spargo, T. (1999). *Foucault and queer theory*. London: Icon Books Ltd.

Stillwagon, R., & Ghaziani, A. (2019). Queer pop-ups: A cultural innovation in urban life. *City & Community*. 18 (3), 874–895.

Trainer, S., Wutich, A., & Brewis, A. (2017). Eating in the panopticon: Surveillance of food and weight before and after bariatric surgery. *Medical Anthropology*. 36 (5), 500–514.

Varangis, E., Lanzieri, N., Hildebrandt, T., & Feldman, M. (2012). Gay male attraction toward muscular men: Does mating context matter? *Body Image*. 9 (2), 270–278.

Walters, T., & Jepson, A.S. (2019). Understanding the nexus of marginalisation and events. In: Walters, T., & Jepson, A.S. (ed.) *Marginalisation and events*. London: Routledge, pp. 1–16.

Wang, C., & Burris, M.A. (1994). Empowerment through photo novella: Portraits of participation. *Health Education Quarterly*. 21 (2), 171–186.

Wang, C., Yi, W.K., Tao, Z.W., & Carovano, K. (1998). Photovoice as a participatory health promotion strategy. *Health Promotion International*. 13 (1), 75–86.

Wong, C.F., Schrager, S.M., Holloway, I.W., Meyer, I.H., & Kipke, M.D. (2014). Minority stress experiences and psychological well-being: The impact of support from and connection to social networks within the Los Angeles house and ball communities. *Prevention Science*. 15 (1), 44–55.

12 The role of events' media narratives in overcoming discrimination, countering negative well-being and . . . effecting social change?

Trudie Walters and Ismail Shaheer

Introduction

Well-being is a complex and multi-layered concept. For example, physical well-being covers one's bodily health, while psychological well-being involves constructs such as autonomy, mastery, purpose in life, personal growth, self-acceptance and positive relations (Ryff & Keyes, 1995). Social well-being includes social coherence, social integration, social contribution and social acceptance (Keyes, 1998), and subjective well-being comprises more nebulous concepts such as life satisfaction and happiness, and being free of excess worry, sadness, anger and guilt (Davidson & Cotter, 1991). Individual and community well-being may therefore be influenced by, and impacted upon, many things – including discrimination and marginalisation (Schmitt *et al.*, 2014; Kauff, Wölfer & Hewstone, 2017; Walters & Jepson, 2019). People may be marginalised for a variety of reasons including disability, race or ethnicity, sexuality, religious beliefs, age or (low) socio-economic status, and the intersectional nature of these factors compounds the effect. Marginalisation renders people powerless and this experience may be both significant and far-reaching. In this context, the media has a major role to play as it both reflects and influences society: it tells stories and project images about who "we" are to others (Karlsen, 2007). However, there are implicit power relations in the telling of media stories – who tells them, what is told, and how it is told (Denzin, 1996). This is especially pertinent for marginalised groups (particularly but not exclusively indigenous and ethnic minority communities) who are often silenced, ignored or represented as the "other" and in such a way as to perpetuate negative stereotypes (Paxton *et al.*, 2011; Olvera, 2016; Allen & Bruce, 2017; Triandafyllidou, 2018).

In a similar manner to media narratives, events can be considered as rich cultural texts that convey and interpret community values and social mores in multiple ways (George, Roberts & Pacella, 2015). Researchers have argued for the transformative effects of event media narratives that counter negative stereotypes and re-centre those at the margins – for both individuals and society (Mackley-Crump & Zemke, 2019). There is a newly emerging body of work being done in this area (see, e.g. Rovito & Giles, 2016; Chen, Mason & Misener, 2018; Walters & Ruwhiu, 2019), and we seek to build on this. Our aim in this chapter is to investigate the possibility that events in, by or for marginalised groups – or, more specifically, the media coverage of these events – may function as conduits to normalise diversity and overcome discrimination, and in so doing counter the negative well-being of such individuals and communities and effect social change. We use the media narratives of a series of seven events around New Zealand, selected as part of a wider study, as the focus of this chapter.

Marginalisation and well-being

There is a significant amount of research at the intersection of marginalisation and well-being, but it is beyond the scope of this chapter to provide comprehensive coverage. Rather, we seek to provide an overview that focuses on areas relevant to our study and gives an insight into the experiences of marginalised groups.

Older people who are marginalised through reduced social networks (through death or relocation), ill health or reduced mobility experience greater loneliness and social isolation, which in turn leads to poor well-being (i.e. physical and mental health) outcomes (WHO, 2002; North & Fiske, 2012; Wood, Jepson & Stadler, 2018; Jepson, 2019). Older women, in particular, feel the marginalising impacts of ageist discrimination: for many, the sense of being "invisible" in the eyes of society can result in low self-esteem and social withdrawal (Clarke & Griffin, 2008; Chrisler, 2011). These experiences may be intensified when poverty or rural/geographically remote living are taken into account, thus highlighting the intersectional nature of marginalisation that exacerbates the problem (Keating, Swindle & Fletcher, 2011). Other ways in which living in rural/remote locations may work to marginalise is through lack of infrastructure (such as schools, banking facilities and reliable broadband connectivity), population aging, outmigration and overreliance on primary industry (Bock, 2016).

For ethnic minority migrants, the (re)settlement experience in a new land can be a mix of homesickness, loss of identity, inability to communicate effectively in a second language, inability to engage in cultural rituals (such as observing seasonal, religious or culinary traditions), inability to engage in meaningful work, and both explicit and implicit racism (Furnham, 1997). Likewise, indigenous communities are also often marginalised through structural inequality and racial discrimination across the justice, education, health and housing systems in countries including the US, the UK, Australia, South Africa and New Zealand (Brennan, 2011; Walker, Spohn & Delone, 2012; Waterworth *et al.*, 2015). For indigenous and ethnic minority children, the after-effects of this discrimination last well into adulthood: for example, increased behavioural problems, loneliness, low self-esteem and lower academic achievement are significant predictors of low-paid and insecure employment, and poorer physical and mental health outcomes later in life. These consequences for well-being may become self-perpetuating intergenerational problems (Chetty, 2014; Priest *et al.*, 2014; Umaña-Taylor, 2016; Kauff, Wölfer & Hewstone, 2017).

It is a similar situation for people with disabilities; being unable to participate fully in everyday life experiences is not only frustrating, but it is also detrimental to their well-being. Exclusionary practices and attitudes in education, work and leisure mean that disabled people face multiple barriers to experiencing a high quality of life. For example, entertainment and business venues are often physically and/or socially inaccessible, airlines may not allow assistance animals, and employers who cannot see past a disability when hiring means disabled people experience higher unemployment/partial employment (Darcy, 2012a, 2012b; Shakespeare, 2014). This discrimination results in a loss of dignity and frequently increased levels of humiliation and anxiety (Darcy, 2012a).

This brief review highlights the fact that marginalised people across the board (whether discriminated against for reasons of age, gender, race/ethnicity, [dis]ability or geographic location) face significant challenges, and these in turn have consequences for their well-being through hindering their economic and employment opportunities, impairing their physical and mental health and constraining their education and academic performance. Such compounding effects of marginalisation frequently have long-term implications for them and their families into the future, perpetuating the effects inter-generationally.

Marginalisation, the media and social change

There is a body of journalism studies literature focused on media constructions and representations of marginalised groups, demonstrating that media narratives exert power through both reflecting and influencing societal perceptions of such groups. Indeed, Wetherell and Potter (1992, p. 84) state that "power develops through 'normalisation', through defining what is usual and habitual and to be expected, as opposed to the deviant and exceptional". The media has the power to silence people or to allow them to speak: what is not reported is as important as what is, and who is absent in the telling of the story is as important as who is present (Denzin, 1996; Thorbjørnsrud & Figenschou, 2014). As noted earlier, negative stereotypes are frequently perpetuated in and by the media. For example, words and descriptors commonly associated with marginalised groups such as Gypsy Travellers, asylum seekers, LGBTIQ+ people, seniors, people of colour (both indigenous and ethnic minority people), the disabled and the fat include: deviant, dirty, violent, dangerous, pitiable, evil, weak and infirm, social welfare fraudsters, "freaks of nature", lazy and lacking in self-control (see e.g. work by Angus & Reeve, 2006; Chesney-Lind & Eliason, 2006; Happer & Philo, 2013; Rimmerman, 2013; Thorbjørnsrud & Figenschou, 2014; Allen & Bruce, 2017; Triandafyllidou, 2018; Crew, 2019; Elliott & Finkel, 2019).

These social constructions play a pivotal role in informing the public about things that they do not have personal knowledge or experience of (such as marginalised people), and in so doing the media has the power to focus public interest and shape public debate (Chesney-Lind & Eliason, 2006; Happer & Philo, 2013; Crew, 2019). For example, increasingly negative media representations of disability in the UK a decade ago resulted in a corresponding hardening of public attitude towards disabled people (Brittain & Beacom, 2016). Furthermore, the marginalised seldom have the power to influence content creation; their voices are therefore silenced, and debates of relevance or interest to them are seldom covered (Denzin, 1996; Allen & Bruce, 2017).

However, research does suggest that the media may have a role to play in driving positive social change and that, with targeted training and support, marginalised groups may be able to make their views more prominent (Ryan, Carragee & Meinhofer, 2001; Happer & Philo, 2013). Recent work by Walters and Ruwhiu (2019) on 15 years of media coverage of an indigenous event in New Zealand (the Māori festival of Puaka Matariki) supports this claim, finding a positive tone that disrupted typically negative media narratives and firmly brought Māori to the centre of attention.

Events, well-being and social change

Critical event studies' scholars are increasingly turning their attention to the nexus of events and well-being, whether explicitly or more obliquely. For example, research has been carried out on events in/for rural and marginalised low socio-economic communities (such as that by Gibson & Connell, 2015; Walters, 2019), refugee and ethnic minority migrants (Hassanli, Walters & Friedmann, 2020; Hassanli, Walters & Williamson, 2020; Walters & Venkatachalam, forthcoming), families (Jepson & Stadler, 2017), the LGBTQI+ community (Mackley-Crump & Zemke, 2019), seniors (Jepson, 2019; Walters & Wright, forthcoming) and youth (Packer & Ballantyne, 2011).

Such research has identified a number of ways in which events work to facilitate physical, social, psychological and/or subjective well-being. These include creating a psychological sense of community, facilitating a means of togetherness in crisis, creating safe spaces for the sharing and expression of identity, overcoming isolation and loneliness, keeping body and mind active, fostering pride and improving quality of life. Indeed, some events may be viewed as sites for resistance that allow marginalised groups to assert control and offer counter-narratives to the ways in which they are represented or to colonise public spaces traditionally claimed by the majority groups in society (Mackley-Crump & Zemke, 2019; Hassanli, Walters & Friedmann, 2020). Referencing Freire's (1970/1996) notion that power imbalances are maintained through antidialogical action, events that encourage the attendance of majority groups may act as a form of dialogue between marginalised and majority populations. Some researchers (see e.g. Duffy, Mair & Waitt, 2019; Hassanli, Walters & Williamson, 2020) have argued that this may help reduce feelings of marginalisation, facilitate acceptance within the wider community and thus contribute to social change. However, this may be undermined by media coverage of the events, which perpetuates *representation of* rather than *dialogue with*.

Based on the above discussions then, we argue that positive media representations of events may be a pivotal factor in normalising diversity and reducing discrimination for marginalised groups. Likewise, negative representations and "Othering" may be detrimental to the achievement of such objectives. As we have already seen, marginalisation and discrimination have a deleterious effect on well-being. Walters and Jepson (2019) assert that events have the power to help cause, perpetuate, resist or overcome marginalisation: in this chapter, we argue that media narratives have a significant role to play in this regard and this, in turn, has implications for individual and community well-being.

Methods

Seven events were selected for inclusion in this research. Each was associated with a marginalised community in New Zealand and included those for/by ethnic minority groups, rural women, disabled people and seniors. Newspaper articles related to each event were sourced from online news database ProQuest, Google News and individual national and regional newspaper websites including New Zealand Herald (www.nzherald.co.nz/), Otago Daily Times (www.odt.co.nz/) and Stuff (www.stuff.co.nz/). A total of 171 relevant articles were included in the analysis (Table 12.1).

Table 12.1 Events selected for the research and the number of newspaper articles collected

Event	*Type of Event*	*Location*	*Est.*	*Host Organisation*	*No. of Newspaper Articles*
Leisure Marching Nationals	Non-competitive sport event for seniors (predominantly women)	Alternates between North/ South Island locations	1991	Individual host teams	13
Pasifika Festival	Pacific Islands cultural festival (food, music, dance, information)	Wellington (New Zealand's capital city)	2006	Wellington City Council	20
Diwali Festival of Light	Indian cultural festival (food, music, dance)	Auckland (New Zealand's largest city)	2010	Various	107
InterACT Disability Arts Festival	Arts event (incl. performing arts, interactive workshop activities, art exhibitions)	Auckland (New Zealand's largest city)	2011	InterACTing Trust (not-for-profit organisation)	9
Holi Festival of Colour	Indian cultural festival (food, music, dance)	Christchurch (the South Island's largest city)	2012	Revel Events (private Indian-owned company)	11
Clutha Women in Business Expo	Business event (B2B, B2C)	Balclutha (small South Island rural town)	2017	Clutha Women in Business Network	3
Diwali Festival of Light	Indian cultural festival (food, music, dance)	Invercargill (small South Island regional city)	2018	Indian Community in Southland (Inc. Society)	8

Critical Discourse Analysis (CDA) was an ideal methodological fit to analyse the data as it studies the written language in relation to its social context (Fairclough, 2013), thereby aligning with the objective of the study. CDA is commonly used in analysing media discourse to understand societal power relations, such as the representation of marginalised communities (Chen, Mason & Misener, 2018). According to Blommaert and Bulcaen (2000), CDA can highlight issues of power asymmetries, manipulation, exploitation and structural inequities in societies, which naturally fits with the context of this study.

The analysis was carried out inductively and followed the three-dimensional framework developed by Fairclough (2013). This framework focuses on analysing the data at three different levels; micro-, meso- and macro-perspectives. Micro-level analysis looks at the written text to understand its meaning. At the meso-level, the focus is on understanding the discourse practice, such as who writes the article, the tone of the article and who the target audience is. Finally, the macro-level perspective explores how the data (i.e. themes identified) fits into the broader societal context.

Findings and discussion

It is important to recognise the power of the media in deciding what events are presented to the public and how to present them (Gist, 1991). The power of the media in bringing the stories about events associated with marginalised communities undoubtedly has implications for how the broader society perceives them and thus for countering negative well-being and effecting social change. The analysis of event media narratives in this study shows that different voices were given the opportunity to communicate their perspectives of the events to the newspaper audience: articles frequently included quotes from political leaders, community leaders of the marginalised communities, general public and event attendees, event participants and members of the marginalised community associated with the event.

While media outlets' justification of what events to cover, how much to cover and what stories to tell can be subjective, the media also often includes stories about events that are already high profile, popular and have commercial implications (Ellman & Germano, 2009). This raises the question of a hierarchy of marginalised communities within a society that has been noted by others (Hassanli, Walters & Friedmann, 2020). For example, here it was observed that the cultural events received more media coverage than the events for older women, rural businesswomen or disabled people, even when the longevity of the events was taken into account (Table 12.1). This is perhaps unsurprising, given that multicultural events are often used by local governments as a part of a strategy to promote tolerance of difference – despite being experienced by some of the mainstream audiences as merely confirming cultural stereotypes through the selective nature of food and entertainment on offer (Duffy, Mair & Waitt, 2019).

Table 12.2 The two principal themes and their contributing subthemes identified during the micro-level analysis

Principal themes	*Sub-themes*
Awareness	Platform to create awareness Subvert stereotypes Preserving self-identity
Acceptance	Significance to community Sense of belonging

We therefore encountered some limitations in our ability to make conclusions about the other (non-cultural) events. Furthermore, while we had hoped to carry out a longitudinal analysis of the data to investigate a change in the media narratives over time, this lack of newspaper coverage was a significant constraint and no conclusions can be drawn in this regard. While disappointing, these observations are telling in terms of Fairclough's (2013) macro-level critical discourse analysis. We argue that they reflect the New Zealand public's perceptions of, interest in, and willingness to accept, stories about ethnic minority groups over those about older women, disabled people and women in business in rural areas.

The micro-level analysis of the media narratives revealed a number of subthemes that were consolidated into two principal themes: awareness and acceptance (Table 12.2). Both of these themes and their contributing subthemes will be unpacked in the remainder of the chapter, using excerpts from the narratives to illustrate the conclusions being drawn.

Awareness

The media commonly referred to the events as an opportunity to increase visibility and create awareness. The article authors and interviewees alike provided similar responses, stating that the events were a valuable platform to educate the wider (mainstream) community:

> This festival is an opportunity that gives these communities the chance to share their own special cultural traditions and foods with the wider Auckland community. (Charmaine Ngarimu, Head of Major Events for Auckland Tourism, Events and Economic Development) – Diwali
>
> (*Scoop*, 13 September 2017)

> It was marketed as: 'An opportunity for all Pacific people to celebrate and share their knowledge, values and beliefs, and supports the vision that Wellington is a vibrant, creative capital'. – Pasifika Festival
>
> (*Dominion Post*, 20 January 2020)

These findings echo previous research that has argued that events can be a crucial platform for creating awareness of and about marginalised communities. For instance, New York City's Pride event (an important event for the LGBTQ+ community), which stemmed from the Stonewall riots in 1969 (McGrody, 2018), is now held annually in many countries (see www.iglta.org/). It is identified as an important platform for raising awareness about different discussions associated with the LGBTQ+ community (Ammaturo, 2016). In the current study, the media are both reporting on the interviewees' views that the value of such events lies in their ability to function as platforms for their marginalised voices and in doing so are, importantly, acting as that platform.

In addition to raising awareness and visibility of marginalised groups, the events served as tools to subvert stereotypes. The newspaper articles reported many interviewees stressing how the events could facilitate a change in perception through exposing mainstream attendees to aspects of the marginalised communities that they had not experienced before:

> Most people think "curry" when they think of Indian food but the Diwali festival is a great opportunity to experience the diversity of Indian cuisine. This is the city's biggest multicultural vegetarian festival.
>
> (Sarah Ell, *New Zealand Herald*, 11 October 2014)

> The highly successful [Clutha Women in Business] event looks set to become an annual fixture, to showcase the entrepreneurial skill of the district's fabulous females.
>
> (Mary-Jo Tohill, *Leader*, 28 September 2017)

> A lot of us have aches and pains but we just get on with it . . . there are teams around New Zealand with some [members] marching with walking sticks. (Val Sherrif, Leisure Marcher and Coach)
>
> (Megan Hunt, *Stuff*, 31 March 2017)

> Mainstream society is sadly ignorant of disabled experiences, and is the poorer for it. Difference and diversity is sidelined, ignored, ridiculed or silenced. Outliers, or marginalised people, often have really valid perspectives because of our objective awareness. (Caitlin Smith, Singer) – InterACT Disability Festival
>
> (*Stuff*, 15 October 2015)

For singer-song writer Natalie Te Pā, the InterACT Disability Festival provides an opportunity for her to counter assumptions about people with a disability through creating a platform for dialogue:

> if I meet someone with assumptions I always get into conversation and tell them more about my story and experience to bust some of those myths about being blind.
>
> (Emily Ford, Stuff, 15 October 2015)

These excerpts support Mackley-Crump and Zemke's (2019) claim that events can help in countering the negative stereotypes attached to minority communities. Similarly, Markwell and Rowe (2003) note how The International Gay Games became a lens to address the "(mis)conceptions of homosexuals" and to "destroy the gay stereotypes" (p. 3), normalising the presence of gay athletes in the sports arena. Here, we see evidence that stereotypes about ethnic minority groups, seniors, rural women and disabled people are being acknowledged, but more importantly, they are also being challenged (at least to some degree) in the media narratives.

The final subtheme was that of preserving self-identity. Marginalised communities who had roots in other countries noted that the events created awareness for them about their own community and identity. Cultural events were thus represented in the media as a useful source of information for attendees to maintain, educate and understand their own culture and identity – even when they considered New Zealand to be home:

> I call New Zealand home . . . but my ethnicity is Indian, that's where my roots are so I think it's dancing and Diwali that has kept me in touch. (Rahul Chopra, festival attendee)
>
> (Hugh Collins, *Stuff*, 11 October 2016)

Having an opportunity to experience their own home culture as a diaspora can help people (re)connect to their roots, and there is an established body of literature dealing with the role of events in this regard. For example, Gsir and Mescoli (2015) report that cultural events can help migrants in understanding their own cultural identity. Laing and Frost (2013) examined Italian-themed festivals in regional Victoria (Australia) and reported that the events were useful in providing a sense of identity for Italian communities in Victoria. Similarly, work by Hassanli, Walters and Friedmann (2020) found that multicultural festivals allowed ethnic minority migrants and refugees the opportunity to express their identity and engage in cultural practices in a safe space. The inclusion of this in the media narratives may help to educate mainstream society about the importance of cultural events for ethnic minority migrant communities.

Acceptance

Attendance and acknowledgement by a country's leaders and other significant public figures within the local population are critical for marginalised communities to feel a sense of belonging (Jalalian Daghigh & Rahim, 2020). Lack of equal engagement of the nation's leaders with minority communities' events may be perceived negatively and signal a preference for some communities over others when considering other opportunities as well (Weiner, 1983).

In this study, many of the newspaper articles reported the attendance of political leaders and high-profile public figures – especially at the cultural events. This has been especially noticeable in more recent years, with their attendance lending credibility to both the events and the communities they represent. This conveyed a strong message to the ethnic minority community and, importantly, mainstream society, that they belong:

> The Indian Festival of Lights kicked off with the lighting of the ceremonial lamp by Prime Minister John Key: 'Your presence means richness in culture, in music, in cuisine . . . we can celebrate our city as the most multicultural this country and one of the most multicultural in the world'.
>
> (Nicole Lawton, *Stuff*, 15 October 2016)

> On October 14, [Prime Minister Jacinda] Ardern joined tens of thousands of people in Aotea Square for the annual Auckland Diwali Festival celebrations.
>
> (*New Zealand Herald*, 19 January 2018)

A number of newspaper articles recognised the importance of diversity within New Zealand communities, framing the contribution of the event as a forum for the mainstream to show their appreciation of such diversity, and thus foster a sense of acceptance and belonging. This was particularly the case not only for the cultural events but also for the InterACT Disability Arts Festival:

> [The Chair of the Indian Student Welfare Association is quoted as saying] Here in Auckland the whole community from different parts of the world celebrates – creating another type of extended family. I love seeing kiwis dressed up, immersing themselves in our culture for a day.
>
> (Alexia Santamaria, *Stuff*, 16 October 2018)

> The Director of the InterACT Disability Arts Festival stressed the importance of the festival as giving recognition to the 20 per cent of the population who identify as being disabled.
>
> (*Auckland Council*, 21 February 2020)

> [The organiser of the Pasifika Festival states that the event is] a great opportunity to bring all those different Pacific people into one place and acknowledge these communities, and appreciate all the colour and flavour and culture they add to Wellington and to our reputation as a cultural capital.
>
> (Andre Chumko, *Stuff*, 20 January 2019)

In addition to showcasing the benefits of diversity, several articles reported the marginalised communities' significance and their contribution to society. The following comments from the media articles presented depict how media reporting of events can establish the importance of marginalised groups in a community:

> Audiences can also learn about the 125-year history of New Zealand's Indian communities at a photography exhibition called Mokaa: the Land

> of Opportunity located in the foyer of [Auckland's premier arts and culture venue] Aotea Centre.
>
> (*TVNZ*, 8 October 2019)

> [The Indian community is] active in business, in government, and many other aspects of Kiwi society.
>
> (Simon Draper, *Stuff*, 23 October 2017)

Findings from this study reflect research undertaken by Walters and Ruwhiu (2019), who reported that media reporting of the Puaka Matariki festival (a festival celebrating the Māori New Year) held in New Zealand was framed in media articles to signify acceptance to the wider non-Maori community. Media reporting of events for marginalised communities can inform and stress their significance to the wider community and how they conform to society's expectations (Buysse & Borcherding, 2010).

Conclusion

In this chapter, we analysed the media narratives around seven events in, by and for marginalised communities in New Zealand, with the aim of investigating whether they may function as conduits to normalise diversity and overcome discrimination, and in so doing counter the negative well-being of such individuals and communities and effect social change. Here we present two major conclusions. The first relates to the limitation we encountered, that of a general lack of media coverage, which functions as a significant finding in itself. Apart from the cultural events (more particularly those in New Zealand's largest cities), the media largely ignored events for the marginalised. This is despite participation of approximately 1,000 women at the annual Leisure Marching Nationals and in excess of 5,000 attendees at the InterACT Disability Arts Festival every year. This, we argue, is an example of the power of the media to effectively silence the voices of the marginalised (senior women, rural businesswomen and disabled people in particular) (Gist, 1991; Allen & Bruce, 2017; Triandafyllidou, 2018).

Where the media *did* cover events, they did so in a manner that (on the surface at least) seemed to promote the "normalisation" of diversity. Narratives included voices of authority such as politicians and community leaders, which lent an air of credibility and acceptance. In addition, the long history and contribution of the Indian community to New Zealand were highlighted in some articles. However, it was disappointing (but not entirely surprising) that the discourse often highlighted "deficiencies" such as including a remark about the bodily aches and pains of some of the senior women, naming the disability, or associating Indian culture with curry. In this way, the subjects of the event narratives are first labelled according to their stereotyped identity. Only then did the narratives go on to subvert those stereotypes using quotes from event participants and attendees to do so, thus giving [controlled, qualified] voice to the marginalised as "overcomers" and superheroes (in the case of the Leisure Marching Nationals and InterACT Disability Arts Festival participants), or as having more depth (in the case of ethnic minority migrants) (Rimmerman, 2013).

We therefore conclude from our empirical sampling that although there is some media coverage of events in, by and for marginalised groups that functions as a platform to raise awareness, it is limited. By not bringing the positive celebrations and achievements of these groups to the attention of the wider public, the media are wielding their power to decide who "counts" and who does not. Through this lack of coverage, then, the media are complicit in maintaining stereotypes of senior women, rural businesswomen and disabled people in particular as incapable of being active, being successful in business, or being creative, respectively.

In closing, then, we return to Wetherell and Potter's (1992, p. 84) statement that "power develops through 'normalisation', through defining what is usual and habitual and to be expected" and find that, for marginalised groups, it is still the invisibility that is usual. And habitual. And to be expected. Their non-newsworthiness renders them invisible, and the opportunity that events provide for the media to help overcome discrimination, counter negative well-being and effect social change is currently being squandered.

Acknowledgement

This study was funded by a University of Otago Research Grant.

References

Allen, J.M., & Bruce, T. (2017). Constructing the other: News media representations of a predominantly 'brown' community in New Zealand. *Pacific Journalism Review*. 23 (1), 225–244.

Ammaturo, F.R. (2016). Spaces of pride: A visual ethnography of gay pride parades in Italy and the United Kingdom. *Social Movement Studies*. 15 (1), 19–40.

Angus, J., & Reeve, P. (2006). Ageism: A threat to 'aging well' in the 21st century. *Journal of Applied Gerontology*. 25 (2), 137–152.

Blommaert, J., & Bulcaen, C. (2000). Critical discourse analysis. *Annual Review of Anthropology*. 29 (1), 447–466.

Bock, B. (2016). Rural marginalisation and the role of social innovation: A turn towards nexogenous development and rural reconnection. *Sociologia Ruralis*. 56 (4), 552–573.

Brennan, E. (2011, May 27). On this day: Indigenous people get citizenship. *Australian Geographic*. Available from www.australiangeographic.com.au/blogs/on-this-day/2011/05/on-this-day-indigenous-people-get-citizenship [Accessed 15 July 2018].

Brittain, I., & Beacom, A. (2016). Leveraging the London 2012 paralympic games: What legacy for disabled people? *Journal of Sport and Social Issues*. 40 (6), 499–521.

Buysse, J.A.M., & Borcherding, B. (2010). Framing gender and disability: A cross-cultural analysis of photographs from the 2008 paralympic games. *International Journal of Sport Communication*. 3 (3), 308–321.

Chen, C., Mason, D.S., & Misener, L. (2018). Exploring media coverage of the 2017 world indigenous nations games and North American Indigenous Games: A critical discourse analysis. *Event Management*. 22, 1009–1025.

Chesney-Lind, M., & Eliason, M. (2006). From invisible to incorrigible: The demonization of marginalized women and girls. *Crime Media Culture*. 2 (1), 29–47.

Chrisler, J.C. (2011). Leaks, lumps, and lines: Stigma and women's bodies. *Psychology of Women Quarterly*. 35 (2), 202–214.

Clarke, L.H., & Griffin, M. (2008). Visible and invisible ageing: Beauty work a response to ageism. *Ageing and Society*. 28 (5), 653–674.

Chetty, R. (2014). Class dismissed? Youth resistance and the politics of race and class in South African education. *South-North Cultural and Media Studies*. 28 (1), 88–102.

Crew, T. (2019). Appleby Fair for all. In: Finkel, R., Sharp, B., & Sweeney, M. (eds.) *Accessibility, inclusion and diversity in critical event studies*. Abingdon: Routledge, pp. 48–64.

Darcy, S. (2012a). (Dis)embodied air travel experiences: Disability, discrimination and the affect of a discontinuous air travel chain. *Journal of Hospitality and Tourism Management*. 19 (1), 91–101.

Darcy, S. (2012b). Disability, access and inclusion in the event industry: A call for inclusive event research. *Event Management*. 16, 259–265.

Davidson, W.B., & Cotter, P.R. (1991). The relationship between sense of community and subjective well-being: A first look. *Journal of Community Psychology*. 19, 246–253.

Denzin, N.K. (1996). More rare air: Michael Jordan on Michael Jordan. *Sociology of Sport Journal*. 13 (4), 319–342.

Duffy, M., Mair, J., & Waitt, G. (2019). Addressing community diversity: The role of the festival encounter. In: Finkel, R., Sharp, B., & Sweeney, M. (eds.) *Accessibility, inclusion and diversity in critical event studies*. Abingdon: Routledge, pp. 9–20.

Elliott, A., & Finkel, R. (2019). Barriers to access: Investigation of plus-size women consumer experiences at fashion events. In: Walters, T., & Jepson, A. (eds.) *Marginalisation and events*. Abingdon: Routledge, pp. 179–192.

Ellman, M., & Germano, F. (2009). What do the papers sell? A model of advertising and media bias. *The Economic Journal*. 119 (537), 680–704.

Fairclough, N. (2013). *Critical discourse analysis: The critical study of language*. Oxford: Routledge.

Freire, P. (1996). *Pedagogy of the oppressed* (Rev. ed.). London: Penguin Books (Original work published 1970).

Furnham, A. (1997). Culture shock, homesickness, and adaptation to a foreign culture. In: van Tilburg, M., & Vingerhoets, A. (eds.) *Psychological aspects of geographical moves: Homesickness and acculturation stress*. Amsterdam: Amsterdam University Press, pp. 17–34.

George, J., Roberts, R., & Pacella, J. (2015). 'Whose festival?' Examining questions of participation, access and ownership in rural festivals. In: Jepson, A., & Clarke, A. (eds.) *Exploring community festivals and events*. London: Routledge, pp. 79–91.

Gibson, C., & Connell, J. (2015). The role of festivals in drought-affected Australian communities. *Event Management*. 19 (4), 445–459.

Gist, M.E. (1991). Minorities in media imagery: A social cognitive perspective on journalistic bias. *Newspaper Research Journal*. 12 (3), 52–63.

Gsir, S., & Mescoli, E. (2015). *Maintaining national culture abroad: Countries of origin, culture and diaspora Robert Schuman centre for advanced studies*. San Domenico di Fiesole: European University Institute.

Happer, C., & Philo, G. (2013). The role of the media in the construction of public belief and social change. *Journal of Social and Political Psychology*. 1 (1), 321–336.

Hassanli, N., Walters, T., & Friedmann, R. (2020). Can cultural festivals function as counterspaces for migrants and refugees? The case of the new beginnings festival in Sydney. *Leisure Studies*. 39 (2), 165–180.

Hassanli, N., Walters, T., & Williamson, J. (2020). 'You feel you're not alone': How multicultural festivals foster social sustainability through multiple psychological sense of community. *Journal of Sustainable Tourism*. DOI: 10.1080/09669582.2020.1797756.

Jalalian Daghigh, A., & Rahim, H.A. (2020). Representation of Muslim minorities in politicians' discourse: Jacinda Ardern vs. Donald Trump. *Journal of Muslim Minority Affairs*. 1, 1–17.

Jepson, A. (2019). 'De-marginalising' marginalised communities: The case of participatory arts events and the over 70s in rural Hertfordshire, the United Kingdom. In: Walters, T., & Jepson, A. (eds.) *Marginalisation and events*. Abingdon: Routledge, pp. 76–92.

Jepson, A., & Stadler, R. (2017). Conceptualising the impact of festival and event attendance upon family quality of life (QOL). *Event Management*. 21, 47–60.

Karlsen, S. (2007). *The music festival as an arena for learning. Festspel in Pite Alvdal and matters of identity*. Luleå, Sweden: Luleå University of Technology.

Kauff, M., Wölfer, R., & Hewstone, M. (2017). Impact of discrimination on health among adolescent immigrant minorities in Europe: The role of perceived discrimination by police and security personnel. *Journal of Social Issues*. 37 (4), 831–851.

Keating, N., Swindle, J., & Fletcher, S. (2011). Aging in rural Canada: A retrospective and review. *Canadian Journal on Aging/La Revue canadienne du vieillissement*. 30 (3), 323–338.

Keyes, C.L.M. (1998). Social well-being. *Social Psychology Quarterly*. 61, 121–140.

Laing, J., & Frost, W. (2013). Food, wine . . . heritage, identity? Two case studies of Italian diaspora festivals in regional Victoria. *Tourism Analysis*. 18 (3), 323–334.

Mackley-Crump, J., & Zemke, K. (2019). The FAFSWAG ball: Event spaces, counter-marginal narratives, and walking queer bodies into the centre. In: Walters, T., & Jepson, A. (eds.) *Marginalisation and events*. Abingdon: Routledge, pp. 95–109.

Markwell, K., & Rowe, D. (2003). The international gay games: Subverting homophobia or selling out? *International Sports Studies*. 25 (2), 5–20.

McGrody, E. (2018). *Pioneers of LGBTQ+ rights: The history of the LGBTQ+ movement*. New York: The Rosen Publishing Group, Inc.

North, M.S., & Fiske, S.T. (2012). An inconvenienced youth? Ageism and its potential intergenerational roots. *Psychological Bulletin*. 138 (5), 982–997.

Olvera, J. (2016). The State, unauthorised Mexican migration, and vulnerability in the workplace. *Sociology Compass*. 10 (2), 132–142.

Packer, J., & Ballantyne, J. (2011). The impact of music festival attendance on young people's psychological and social well-being. *Psychology of Music*. 39 (2), 164–181.

Paxton, G., Smith, N., Win, A.K., Mulholland, N., & Hood, S. (2011). *Refugee status report: A report on how refugee children and young people in Victoria are faring*. Melbourne: Victorian Department of Education and Early Childhood Development.

Priest, N., Perry, R., Ferdinand, A., Paradies, Y., & Kelaher, M. (2014). Experiences of racism, racial/ethnic attitudes, motivated fairness and mental health outcomes among primary and secondary school students. *Journal of Youth & Adolescence*. 14, 1672–1687.

Rimmerman, A. (2013). *Social inclusion of people with disabilities: National and international perspectives*. New York: Cambridge University Press.

Rovito, A., & Giles, A.R. (2016). Outside looking in: Resisting colonial discourses of Aboriginality. *Leisure Sciences*. 38 (1), 1–16.

Ryan, C., Carragee, K.M., & Meinhofer, W. (2001). Theory into practice: Framing, the news media, and collective action. *Journal of Broadcasting & Electronic Media*. 45 (1), 175–182.

Ryff, C.D., & Keyes, C.L.M. (1995). The structure of psychological well-being revisited. *Journal of Personality and Social Psychology*. 69, 719–727.

Schmitt, M.T., Branscombe, N.R., Postmes, T., & Garcia, A. (2014). The consequences of perceived discrimination for psychological well-being: A meta-analytic review. *Psychological Bulletin*. 140 (4), 921–948.

Shakespeare, T. (2014). *Disability rights and wrongs revisited* (2nd ed.). London and New York: Routledge.

Thorbjørnsrud, K., & Figenschou, T.U. (2014). Do marginalized sources matter? A comparative analysis of irregular migrant voice in Western media. *Journalism Studies*. 17 (3), 337–355.

Triandafyllidou, A. (2018). A 'refugee crisis' unfolding: 'Real' events and their interpretation in media and political debates. *Journal of Immigrant & Refugee Studies*. 16 (1/2), 198–216.

Umaña-Taylor, A.J. (2016). A post-racial society in which ethnic-racial discrimination still exists and has significant consequences for youths' adjustment. *Current Directions in Psychological Science*. 25 (2), 111–118.

Walker, S., Spohn, C., & Delone, M. (2012). *The colour of justice: Race, ethnicity and crime in America* (5th ed.). Wadsworth: Cengage Learning.

Walters, T. (2019). 'Proud to be South D': Perceptions of a street festival in a marginalised community in New Zealand. In: Walters, T., & Jepson, A. (eds.) *Marginalisation and events*. Abingdon: Routledge, pp. 36–54.

Walters, T., & Jepson, A. (2019). Understanding the nexus of marginalisation and events. In: Walters, T., & Jepson, A. (eds.) *Marginalisation and events*. Abingdon: Routledge, pp. 1–16.

Walters, T., & Ruwhiu, D. (2019). Navigating by the stars: A critical analysis of Indigenous events as constellations of decolonization. *Annals of Leisure Research*. DOI: 10.1080/11745398.2019.1643749.

Walters, T., & Venkatachalam, T.S. (forthcoming). The difference Diwali makes: Understanding the contribution of a cultural event to subjective well-being for ethnic minority communities. *Event Management*.

Walters, T., & Wright, R. (forthcoming). Marching out of the margins: The health and well-being benefits of active ageing through participation in an annual sports event. In: Sturm, D., & Kerr, R. (eds.) *Sport in Aotearoa/New Zealand: Contested terrain*. Abingdon: Routledge.

Waterworth, P., Pescud, M., Braham, R., Dimmock, J., & Rosenberg, M. (2015). Factors influencing the health behaviour of indigenous Australians: Perspectives from support people. *PLoS One*. 10 (11). DOI: 10.1371/journal.pone.0142323.

Weiner, M. (1983). The political consequences of preferential policies: A comparative perspective. *Comparative Politics*. 16 (1), 35–52.

Wetherell, M., & Potter, J. (1992). *Mapping the language of racism*. New York: Columbia University Press.

Wood, E., Jepson, A., & Stadler, R. (2018). Understanding the well-being potential of participatory arts events for the over 70s: A conceptual framework and research agenda. *Event Management*. 22 (6), 1083–1101.

World Health Organization (WHO). (2002). Active ageing: A policy framework. Available from www.who.int/ageing/publications/active_ageing/en/ [Accessed 13 May 2020].

13 Touring academic events with a tiny house "conference fringe"

Artistic welcome in a mobile storytelling shed as relational research into invisibility and (non-)belonging

Christian H. Hanser

Introduction

Academic conferences have been identified as crucial gateways for researchers' career advancement. Academia's mobility imperative (Henderson & Moreau, 2020) leads to increased efforts of individuals to travel and become visible to conference networks with the hope to belong. At the same time, diverse forms of marginalisation are perpetuated during such events. Top–down hierarchies are easily enacted rather than profoundly challenged: "significant issues remain around accessibility, diversity and inclusion for many groups including women, academics with disabilities and impairments, students, early career and precarious academics, and those with caring responsibilities" (Walters, 2019, p. 230). Studies increasingly point at ways how invisibilities and discriminations trouble conventional inclusion logistics (Brown, Thompson & Leigh, 2018; Dashper & Finkel, 2020) and how the subjective experience of non-belonging significantly influences many areas of life, particularly the social and cultural dimensions of well-being (Gillberg, 2020). To move beyond the academic sector's high tension, its precarity, lack of diversity and its structural and endemic shift of systemic pressures onto the individual's mental health (Shahjahan, 2019), well-being shall not be defined through dominant deficit-models of health. Instead of zooming into knowledge about disease through statistically measurable indicators, an emancipatory focus transcends the reliance on ill-health. The search for subjective well-being and existential coherence asks how health and well-being manifests and how it can be facilitated amidst and *despite* many negative, disabling factors (Mittelmark & Bauer, 2016).

Sense of belonging and purpose at academic events is not automatically guaranteed once a presentation has been delivered to a large audience. Academic conferences frequently fall short of their inclusivity potential, lacking gestures of welcome on an existential and political dimension. Conferences have been more thoroughly scrutinised through "a tripartite approach to considering accessibility in [academic] conference organisation: physical accessibility, financial accessibility and cognitive accessibility" (Walters, 2019, p. 230). Relational inquiry as a researcher stance for event engagement looks beyond the measurable to include other factors that can impede attendees from meaningfully engaging in activities despite physically being in the same venue. Highly complex and difficult to categorise, well-being is then a fluid element in a wider human ecology where issues of social justice as well as the intrinsic struggle for a sense of coherence connect, fold into each other and challenge the neat definitions of well-being in the biomedical tradition, perpetuating "the transformation of everyday human conditions and behaviours into diagnosable, treatable disorders" (Russell, 2020, p. 296). Researching well-being then becomes not only a messy but also emancipatory task.

This chapter connects to qualitative discussions on engagement and transformation as a positionality in research where the separation lines between "researcher" and "researched" can be constructively blurred to make space for "liquid knowing" (Nelson, 2013) as well as for lived experience (Brown & Leigh, 2020). Challenging the traditional scientific mindset of detachment and distance, this chapter emerges while I am personally navigating the contradictions of academia as a PhD student. My encounter with a potentially enabling as well as, in parallel ways, profoundly disabling work environment has nurtured an in-between stance.

I operationalised a "safe space" itinerary for welcoming delegate interactions and networking outdoors and beyond cognitive overstimulation and the spotlight of exposure at the centre. This chapter opens methodological *and* practical spaces of mobility and stillness in which data on academic events are not observed, documented and word-processed but actively disrupted by engaged research processes. This experiment seeks to craft sites of dignity, authenticity and hospitality as participatory forms of existential care. The travelling tiny house conference fringe explores formats through which individuals in the knowledge economies could be enabled to find their own sense of place, purpose and positionality, recognised in their valuable difference and singularity. A wider question this chapter then seeks to explore is how the literature on marginalisation in Critical Event Studies can be playfully and disruptively applied to allow "event-related experiences of individuals and communities, existing 'at the margins' for reasons of age, religious beliefs, sexuality, race and ethnicity, socio-economic status, disabilities, refugee status and geographical location" (Antchak, 2020, p. 264) to be given the free-roaming space to experiment micro-formats for resilience *and* resistance.

After discussing the context of using a tiny house, a conferencing autoethnography informed by my perceptions from the outskirts of sociability norms of large-scale events narrates how a mainstream conferencing format can trigger information overload and a silent wish to withdraw. The crafting of a participatory conference locality then allowed to attend academic events on alternative, more hospitable and intrinsically more meaningful terms. I discuss some implications for future event practices and methodological processes.

Framing the fringe: an itinerant events venue

The methodological processes narrated in this chapter relied to a large part on the possibility to make use of my own tiny house as a public encounter space. This pre-existing support infrastructure that I brought into my research design

- covered complex event logistics (through its roadworthiness, reduced costs of installation and acquisition of materials)

- transferred years of experience from non-academic events into the research process (having used the Welcome Hut intervention model as a pop-up pedagogy in my role as community facilitator)
- contributed a symbolic dimension and public visibility (with the use of evocative, artistic communicative material that attracted attention beyond text – see Figure 13.1)

The tiny house is my departure point to research belongingness in the proposed way. Still, the ideas and standpoints for this emerging methodology do not rely on this specific shepherd's hut venue to be further experimented elsewhere. The conference tour with the vagabond encounter space opened up new possibilities for methodological co-creation. The autoethnography at the start of this chapter therefore gradually makes way for a long-term perspective of collective conference fringes and leads to the possibility of (early career researchers') conference

Figure 13.1 Artwork from the Welcome Hut project

Source: Reproduced with permission: Freya Payne.

commons for a post-COVID-19 future of more-than-virtual networking. The reliance on the specific Welcome Hut initiated by the author in 2010 and its corresponding logistical, pedagogical and symbolic support could be seen as an open question to other researchers who want to broaden spatial dimensions of academic investigations: what other sites of encounter and physical designs could we create and then bring into research processes?

From 2018 to 2020, the Welcome Hut was housed by Moray House School of Education and Sport at the University of Edinburgh as a novel outdoor classroom in the School's quad (see Figure 13.2) for the purpose of a PhD project in educational policy on outdoor education and Initial Teacher Education curricular reform. In an unprecedented and experimental presence of the mobile shed on the UK campus sites, the Welcome Hut UK conference tour has allowed me to observe and apply the potential of a mobile tiny house methodology. While the actual PhD is not focused on Critical Event Studies, I had received an enthusiastic response from my doctoral supervisors about the idea to pursue an immersive events-driven first-year literature review.

With a background in years of itinerant educational practice and activism, my return into academia meant to work towards a meaningful valorisation of "peripheral" knowledges. Instead of only identifying the published literature in expert journals for the exploratory phase of my doctoral thesis, my aim for bottom–up

Figure 13.2 The Welcome Hut's permanent location in the Moray House quad, Edinburgh
Source: Author.

educational policymaking was to find ways of being in contact with less visible knowledges as well as non-academic, public debates as early as possible in the PhD: on the one hand, "exploring alternative scientific practices made visible through plural epistemologies of scientific practices and, on the other, . . . promoting interdependence between scientific and non-scientific knowledges" (Santos, 2012, pp. 57–58). While also making use of the university library, I identified academic conferences in the social sciences with a likely presence of stakeholders beyond the academy: representatives from activism, diverse professional fields from health to design, regional partnerships, policy, media and citizen science. I applied for these conferences with a suggestion for outdoor workshops in my experiential tiny house "dream tank". All such proposals got accepted through peer review, one proposal received a "best abstract" conference award.

My PhD conference tour intended to challenge the dominant model of the first-year researcher's literature review about the researched and to investigate voice at the margins of conference spotlights. My immersive events methodology emerged almost "by accident" out of what was meant to be a public engagement exploratory tour to get my PhD research started. The free-roaming nurturing of a conference fringe soon developed a very intriguing methodological life of its own. This did not distract from the work on my PhD but allowed my own doctoral training phase to be significantly led by dialogical encounter instead of the often mentioned solitary "dungeons" of grad school. PhDs can find depth and originality in visceral experience.

Calling my experiment a "conference fringe" is equally not the result of an inward-looking, desk-based trajectory but an inspiration coming from the reciprocity of public engagement. This term was coined by one of the first visitors to the Welcome Hut at the first conference appearance during my PhD. An inhabitant of the city of Edinburgh – a dog walker who did not visit the conference venue as an academic delegate – was the first to approach the public hut and, in hearing my explanations for this unusual installation, concluded that I was creating a "conference fringe" just like the City of Edinburgh festivals also have established their own largely outdoors Fringe festival. The arts of encounter drive this methodology more than the vocabulary of academic categorisation. A timidly conceptualised "fringe" carries a deliberately fluid and unoccupied meaning. In this arts-informed spirit, the tiny house conference fringe is an accidental research methodology, harbouring improvisation and unexpected encounters to move beyond a hierarchy of knowledges that is cemented by linear research as mastery and data management.

I highlight the following academic events which I attended as a PhD student where the conference fringe had been facilitated in a continuous presence:

1 European Congress of Qualitative Inquiry, February 2019, University of Edinburgh. Theme: Qualitative Inquiry as activism (Welcome Hut Gamechanger Panel delegate workshops then changed to a drop-in continuous presence during last day)

   350–400 delegates/18 visitors at the hut

2 UNESCO Chair of Refugee Integration through Languages and the Arts Spring School, May 2019, Heart of Scotstoun Community Centre Glasgow. Theme: The Arts of Integrating (Welcome Hut continuous presence for two days)
140 delegates/35 visitors at the hut
3 World Community Development Conference, June 2019, University of Dundee. Theme: People, Place and power (Welcome Hut continuous presence for one day and delegate workshops)
500 delegates/40 visitors at the hut
4 Centre for Child Well-being and Protection Conference, September 2019, University of Stirling. Theme: Children's and Young People's Mental Health and Well-being: Communities, families, resilience and resistance (Welcome Hut continuous presence and delegate workshop for one day)
95 delegates/30 visitors at the hut
5 Scottish Educational Research Association Annual Conference, November 2019, University of Edinburgh. Theme: Creative visions for educational futures (Welcome Hut continuous presence for two days)
170 delegates/40 visitors at the hut

This Scottish campus tour was concluded as planned in November 2019, shortly before the unexpected cancellations and postponements of in-person conferences due to COVID-19.

Illustrations of some of the physical setting of the tiny house installation (see Figures 13.3, 13.4 and 13.5) and encounters (see Figures 13.6 and 13.7) are selected below, a representation that is inevitably incomplete because a seamless documentation of the experiment had not been the purpose of this relational inquiry. The actual conference fringe experience depended on a multitude of factors: conference size, theme, fees, location of fringe and main event, schedules and their communication, the distance between the main event venue and the outdoor hut, the conference flows and arrival/departure movements, the varied Scottish weather conditions, etc. Instead of evaluating comparative data sets of conference contexts that even proved to be completely different from one day to the next within an event, I adopt the lens of incomparability, thus shifting the different experiences towards the singular rather than generalisation:

> singularities demand a relatively undisciplined kind of disciplinary knowledge production, not in the sense of unrigorous work, but in the sense of a careful and active work that, quoting Richa Nagar (2014, p. 2), 'places question marks on the utility and logic of neat positions and categories', given that those very positions and categories reflect the epistemic hierarchies, logic and investments of our own locations as EuroAmerican geographers. Singularities demand to be read in their specificity.
>
> (Jazeel, 2019, p. 12)

Instead of schematising what happened at the aforementioned conferences, another question appears much more important in what I consider a research design whose

Figure 13.3 Hut and hammocks behind the main conference building, Dundee

Source: Author.

Figure 13.4 Investing a green area next to the conference building, Stirling

Source: Author.

Figure 13.5 The fringe in the Moray House quad for the final conference of the tour, Edinburgh

Source: Author.

main ambition is not data gathering, order and protocol but fragments of caring which unfold their own lives. The wooden hut with its wood fire stove primarily seeks to facilitate change emerging from existential care as an act of hospitality.

The conference fringe hosted open-ended, improvised storytelling as an inclusive gesture among conference delegates, with a focus on temporality in the first installation (Hanser, 2020a). At all events, many delegates entered the hut because of cognitive overstimulation at the end of the day: a wood fire stove instead of *Death by Powerpoint* at 5 pm. Campus catering or administration staff were

Figure 13.6 Quiet time for delegates when my dog is allowed to join the conference fringe, Stirling

Source: Author.

Figure 13.7 Encounters during a workshop for informal storytelling inside the "dream tank"

Source: Author.

invited to sit down with academics who took refuge after periods of constant networking. Some decided to deliberately miss out on keynotes or workshops and simply wanted to catch their breath in an environment that took a lot of energy to attend. Others felt the presence of my dog around the hut, if permitted by the venue's rules, as soothing. If they were entering the hut to sit next to the wood fire stove, delegates were invited to shift personal narratives towards hopes, dreams and imagining new ways to find meaning in their academic work environment. As a non-clinical space for collective artistic expression, delegates could go with their own intrinsic flows. I had previously facilitated the experience of "introvert conviviality" in diverse professional contexts (Hanser, 2020b). The Welcome Hut project had been designed as a civic sanctuary, open particularly to those incapable or unwilling to perform an ideal of able-bodied extraversion in a rarely introspective society. The values carried through the itinerant conference fringe are embedded in the personal life experience of exclusion and marginalisation which frequently risk being interpreted as an individual's withdrawal. Such lived experiences reveal the problematic dogma of socialising norms that are perpetuated in normative design but are increasingly challenged by research on ableism.

Introvert conviviality: an events autoethnography from the in- to the outdoors and back

As a postgraduate student in 2008, I had signed up for many conferences and academic networking events proposed at universities in the UK and in continental Europe. Instead of helping me to pursue a career in research, my experience of attending conferences accelerated the decision to drop my academic ambitions and move on elsewhere – despite sincerely enjoying the spirit of inquiry. I recall research events as profoundly disabling and dislocating. Instead of allowing newcomers to integrate, I often had the feeling that they were required to be showcases for established research. Academic event structures such as keynote status and symposia rewarded outgoing, eloquent sociability to be consumed as a spectator. Conferences comforted those who were already "in" and preferably well advanced in the game. I came to understand the meaning of networking as assimilationist pressure: "assimilation always implies coming into the game after it has already begun, after the rules and standards have already been set" (Young, 1990, p. 164). Entering a mainstream university conference felt like witnessing an exercise of knowledge territorialisation, not the decolonisation towards novel (a-)venues of thought. Conferences made me want to hide rather than network. I stood in the corners and watched the eloquent speak. The rituals on centre stage do not necessarily reward depth:

> For the presenter, it is then virtually impossible to say anything much to an audience who will almost certainly not have read anything in advance. Papers hence become brief and bland, simply to ensure their digestibility in such inhospitable contexts. And that is just as well, because given the necessity to attend a lot of conferences and to present a paper there to get your expenses

> reimbursed, most delegates probably do not have a lot of groundbreaking findings or arguments to report anyway.
>
> (Parker & Weik, 2014, p. 175)

Instead of helping to develop a sense of place in the contested territories of science, conferences appeared to act as "spaces in which solitude is experienced as an overburdening or emptying of individuality" (Augé, 1995, p. 87). A mainstreamed delegate "obeys the same code as others, receives the same messages, responds to the same entreaties. The space of non-place creates neither singular identity nor relations; only solitude, and similitude" (Augé, 1995, p. 103). Academic conferences as non-places carry their own attractiveness: they can offer in-betweenness and transit anonymity as a welcome travel break with career repercussions. As one can disappear in the massification of large-scale networks, the solitary traveller who has found stability in the academy is not disadvantaged in this system. While some can drift pleasantly in the convenient anonymity, such a stroll cannot be performed by all delegates alike. Attending such events had little resemblance to what I would have done in my own leisure time. Some may enjoy the spotlights of networking, others may find pleasure in attending without encounter. Yet others will fatigue from the noises of constant exposure. The taken-for-granted sensory participation in the non-stop flooding from stimuli during a packed, breakfast-to-dinner conference day blocked out any possibility of belonging. There was no hiding-away to digest inputs in events that were celebrating the ability of full-time presence over fatigue.

I had been looking for reciprocity away from the podium. However, top–down academies "teach how to speak but not how to listen. At most they may teach how to hear, but not how to listen" (Santos, 2018, p. 176). Academic events lacked connectivity, materiality, bodily dimensions of learning, sensory hubs and appeared blind and unreceptive to the poetry of place so closely at hand at each singular conference venue. At that time, I wondered if something might be wrong with me, not with the conferences that I was attending. My engagement in research silently faded out. I changed my professional orientation to the tertiary sector and a more horizontal, less "spectacular" idea of events. Instead of freezing in the non-places of academia, I looked for meaningful environments with self-regulated dosage of input and data.

Only after having worked within some collaborative interagency, transdisciplinary event organisation teams in tourism, community development and the arts, I was able to put my academic alienation into perspective. Non-belonging often reflects overarching values of events which are played out in hierarchical ways. Change requires a deeper rethinking of the power structures of event formats:

> events may help cause marginalisation, they may help perpetuate it, they may help resist it or they may help overcome it. These outcomes are an outward manifestation of power relationships and whether event managers choose to adopt inclusionary or exclusionary approaches to event organisation.
>
> (Walters & Jepson, 2019, p. 10)

Preformatted events then appear to be conceived profoundly reductionist to human and more-than-human ecologies. In almost a decade away from linear academic career trajectories, I had sketched out and crafted a personal answer to an existential question for an ex-academic struggling with belonging: how would my own conference/event/public space ideally look like? In 2008, I had participated *indoors* in academic events with the anxious tendency to leave, always sitting in the back of seminar rooms as close as possible to the exit doors. In 2018, at the start of the *outdoors* tiny house conference tour, there was no longer the sudden urge to flee and find a safe spot. I had already learned to sense settings and environments as enabling or disabling. I had internalised that my vulnerability wasn't the problem. My own tendency to hide had become a competency as gateway to opening up a fringe for a diversity of cognitive standpoints among delegates. The positionality that Hill Collins has described as the standpoint of "outsider within" (1999) now resonates existentially with me and encourages me to make creative and integrative use of an autobiographical space. The civic sanctuary operating as an informal drop-in venue offers shelter from standardised ways of being sociable in public. In its public sphere visibility, it exposed many delegates to an alternative to able-bodied extroversion. A calm, tiny and cosy form of conferencing is possible. Fatigue is an inevitable part of the precarities of Early Career Research. It had however been much less exhausting to run the conference fringe by myself, to welcome delegates in my hut than be exposed to the information overload and overstimulation of the indoors keynote venue engineered for a maximised standard of exposure and spectacle.

Most importantly, I discovered that many other delegates face similar struggles with conference designs and also miss the kind of academia in which they can just be different and still belong to their professional networking contexts without suffering through days of excessive cognitive load and fatigue. The idea of quiet rooms has now made its way into event planning inventories, and there are many reasons for thinking that a rather "odd" sensory fatigue is a widely experienced but rarely talked-about phenomenon: "Inside the room were several comfy sofas so that people could lie down for a rest or even a nap. . . . The space was used throughout the day by participants who were overwhelmed, who needed to withdraw, who needed a quiet space and those who needed to regroup. Conferences are by their nature, busy and overwhelming. The quiet room allowed people to dip in and dip out again, to rest, and to return" (Brown, Thompson & Leigh, 2018, p. 84). Silence and listening can sit independently right next to speech.

Ableism and quietly enabling spaces

The appearance of quiet rooms at conferences allows a sense of connectivity and community without prescribing constant exposure, yet recognising the deep relationality inherent in timid, unvocal existential struggles of individuals: "Ableism inside academia is nothing less than a lack or absence of recognition. This is a structural problem that cannot be reasonably resolved at the individual level, though it is at the individual level that the implications and consequences of ableism are felt most keenly, and already precarious work conditions easily become untenable" (Gillberg, 2020, p. 19). Academic institutionalised events are frequently experienced as exclusionary in tacit struggles for belonging:

> Ableism is very much ingrained in the social context. Indeed, ableist attitudes are so deeply rooted within ourselves that we internalise certain ways of working and expect ourselves as well as others to meet these requirements, with some potential adjustments to 'close the gap' and 'make up for the shortfalls'. Any disability, chronic illness or neurodiversity is seen as a deficit. In reality, adjustments are attempts to provide quick fixes for more systemic problems and concerns around accessibility in academia.
>
> (Brown, Thompson & Leigh, 2018, p. 83)

Belonging and connection are mediated by disabling or enabling spaces, not only by an individual's capacity to process the intensities of an overwhelming social setting. In diversity-averse social spheres, "distress and disablement are the product of wider social factors more than individual impairments" (Chapman, 2020, p. 57). A disengaging conference experience is not something that needs to be "fixed" by the person going into anxious hiding. Networking ought to make space for a positive recognition of socially produced invisibilities. Conradson (2016) refers to a study of Horton and Tucker (2014) illustrating this invisibility of otherness:

> amongst faculty and staff who identified as having a "mental health condition," 53% had never disclosed this to anyone in their workplace. Such a finding appears to reflect some notion of a "superableist geographer," a form of academic subjectivity in which the "performative capacities of self-confidence, productivity, stamina, ambition and charisma were anticipated and valorized".
>
> (p. 242)

The divergent standpoints assembled in neurodiversity, particularly from an angle informed by (architectural) design and human geography (Hall & Wilton, 2016; Judge, 2018; Hughes, Franz & Willis, 2019), can then point out the obstacles as well as enablers in a massified events sector that is not always able in its ableist agenda to propose inclusive, emancipatory and relational spaces to breathe. There is a distinction to be made between event design which sees inclusivity as an add-on for individuals to fit into already established frameworks, and event design which is anchored in the transformation of disabling social settings for the silent others. Neurodivergent experimentations come from powerful own standpoints of lived experience which – if given the space to unfold – can develop an own centrality of differently open doors.

Researching differently the invisibilities of (non-)belonging

How can the existential dimension of well-being be witnessed, not scientifically colonised? The Welcome Hut conference tour deliberately invested the outskirts and not the spotlight visibility often aspired to in large-scale conference programme. This fringe positionality seeks to respond to a methodological challenge: researching invisibilities while knowing that the neoliberal academic code rewards constant visibility and exposure (Nicolson, 2017). The choice of methods and the own stance of the researcher are intertwined and can be explored by investigating the way researchers have personally made sense of invisibility. With the intention to research the fragile existential dimension of well-being and events, only certain scientific stances help in approaching the invisibilities of non-belonging and alienation in non-intrusive ways. Modern post-enlightenment science's relationship to invisibility is highly ambivalent and can be considered a blind spot of linear and extractivist modes of research:

> Modern science relies on the assumption that the invisible is more important than the visible and that, therefore, science's mission is to disclose and unveil, in the sense of uncovering. Paradoxically, this assumption is grounded on a meta-assumption that contradicts it. . . . [S]cience is interested in the invisible that can be made visible by scientific work, not in the invisible that science itself generates.
>
> (Santos, 2018, p. 172)

Critical research methodologies tend to detect exclusions, but participatory and engaged forms of inquiry are necessary to address a type of marginalisation which is not expressed and made visible from a detached observer's point of view. This requires a significant shift in research motivations. A researcher confronted with the invisibilities generated by science itself cannot make order in the dark. Science cannot point all flashlights into the direction of this uncertain element and flood this blind spot with the own methodological instruments of lighting. The risk here is to expose, to use methodological tools to master and control, and ultimately to colonise this scientific space of existential (non-)belongings.

Another reaction to invisible alienations could be to embrace this fragile unknown. Having navigated within these precarious landscapes of invisibility and marginalisation allows to embrace the fragility of not knowing, of not controlling a research process and instead drifting with the fluidity of the fringe. Embracing the dark of the periphery is however not about descending into the abyss of the so-called margins. Wandering with the fringe allows to switch off the floodlights of scientific mastery and to gradually acknowledge the dimmed light of the peripheries. Listening to the languages of non-belonging refers to research as caring and witnessing rather than colonising through coding. The fundamental difference between these two approaches is that in the one, the researcher assumes to be bringing light into the alleged dark (and historically invites himself as an expert). In the other, non-interventionist model of research as co-presence, the researcher hopes to be invited as a temporary visitor into life worlds to perceive the existing lights and learn from the very own lanterns of the fringes.

The conference fringe deliberately opens a space of inquiry for the invisibilities generated by science itself: non-validated experience, emotion and the existential meaning of fragility. As such, the conference fringe is an arts-based intervention for pop-up encounters more than a research methodology pursuing positivist approval. The adopted stance is anchored in epistemic humility, "the realisation that our knowledge is always provisional and incomplete" (Angner, 2020, p. 1). With opting for the scientifically obscure invisibilities to be researched comes a specific choice of methodological instruments. Methods can shed light, but if applied in unfitting contexts, they can also cover up the dark. Relational research into marginalisation and invisible dimensions of well-being embraces the liminalities of the dark. Event research seeks to challenge hidden exclusions (Walters & Jepson, 2019). But even in the recent literature of Critical Event Studies, there is a risk to privilege critical writing over the emancipatory transformations of conditions: to uncover rather than to unleash. Uncovering allows to critique instrumental and managerial events through observational rigour but stays too comfortably in its own detached, scientific mindset: "*rejecting* commodification is not the same as *resisting* it" (Spracklen & Lamond, 2016, p. 87). Unleashing implies giving up some of the control that scientists conventionally hold. Paradoxically, uncovering can lead to "covering up."

The proposed methodology is one of opening up rather than narrowing down and occupying. It is incomplete and messy and reflects the impossibility of forced accuracy. In this move from data gathering to the listening of encounters, from expert-leading to facilitation and from rigid protocol to improvised care, not knowing and not necessarily having to know everything is the researcher's stance that makes space for other knowledges and integrative ecologies of being.

A methodology without protocol: allowing the unexpected

What actually happened inside and around the conference fringe has somewhat remained a mystery until now. A systematic data collection was not intended. None of the stories of the fringe's visitors had been recorded on audio or video. As an intimate space seeking to propose an alternative to the perceived audit culture of academia, not documenting the improvised flows was a question of trust as well as of research ethics that relied on fluidity in a drop-in venue. Conference delegates were encouraged to take ownership of a space for *being* rather than doing, an interlude of introspective stillness within often hyperactive conference flows:

> Affects such as stillness can be induced through particular forms of individual and collective practice. This is about the orchestration of feeling, in terms of the capacity of actors to summon, modulate and sustain certain kinds of affective state and field. Examples of such affective choreography are all around

> us [. . .] For while it may be difficult to alter the world beyond ourselves, we retain some capacity to choose our emotional and cognitive response to it.
>
> (Conradson, 2011, p. 73)

This research methodology is then less about retrospectively asking what exactly happened each time a delegate entered the tiny house. Instead, research becomes a gateway towards forward-looking and co-constructed transformational narratives by asking how such affective and existential hubs for imaginaries can be facilitated beyond rigidities of cognitive overload.

This approach to research demands the scientific humility of accepting that the experience cannot be fully explored and can still reveal insights by staying to a large extent invisible:

> The challenge is to research with modesty and attention, which requires being open to following different frictions, traces and meanings, fragments and shards that provoke us in all kinds of settings . . . the point about not subsuming knowledge is a challenge of then remaining with – abiding by – that which surfaces as 'unfathomable' and remains a trace or fragment, but which is nonetheless productive in how it unsettles concepts, positions, and arguments, and discloses a world composed of distinct kinds of difference and resonance.
>
> (McFarlane, 2019, p. 31)

The researcher becomes an attuned carer. In this context, the absence of data allows the researcher to perceive the presence of something more valuable for existential investigation. Instead of extracting findings from encounters, the conference fringe has opened a participatory space to sketch out ways of caring and listening and to experiment with different existential imaginations of being an academic in a different way. The failure to produce a comparative action plan is at the same time the achievement of keeping a site of imagination unoccupied and as little as possible colonised by outcomes-driven agendas. Instead of maintaining an individualistic bubble, collective conferencing cultures were cultivated with care and time:

> Often, the storyteller wants a listener/analyst/researcher to "get into" his or her story, whereas a story analyst, especially a researcher, may be centrally interested in what he or she can "get out" or "take away" from a story (Greenspan, 1998). We see a world of difference between treating stories as "data" for analysis – thus privileging the standpoint of the analyst – and encountering stories experientially – thus privileging the standpoint of the storyteller. In the former case, how a story makes sense is strictly a scientific/analytic question; in the latter case, it's an ethical and relational one . . . in the latter instance, storytelling is a means of being with others, of thinking with their stories in order to understand and care for them.
>
> (Bochner & Riggs, 2014, p. 16)

There is a danger that recruiting diverse publics to have them physically attend an event transforms into tokenism. Such attendance imperatives are opportunities to be more relationally sculptured with meaning. My approach seeks to shift the focus to the existential well-being that manifests when meaningful sites for *being* (which includes being fragile and/or divergent in perception and networking habits) are opened up amidst the controlled and conditional visibilities of standardised event formats. The crafted space is sheltered in its events-affiliated exposure and exposed in its aim as a *civic* sanctuary (see Figures 13.8 and 13.9).

Figure 13.8 The hut's entrance door – the mystery of what happens inside, Stirling

Source: Author.

Figure 13.9 The hut installed on a campus – exposed and sheltered at the same time

Source: Reproduced with permission: Mihaela Bodlovic, 2019 Festival of Creative Learning.

Different forms of art could be combined with my initial narrative inquiry of the conference fringe experience. The hut offered a platform for artistic, visual expression such as drawing, participatory poetry and mapping to leave a trace. The scope of artistic expressions could be enlarged in a future conference fringe that could be run by a team of hosts rather than an individual researcher host. Besides my own researcher diary, there are many story fragments. Many interactions with conference organisers were documented as well as the written feedback from conference delegates anonymously highlighting their experience in the hut on the evaluation sheets for the main event. Delegates were also highly active in tweeting about their experience in the hut and created digitalised perspectives from the fringe.

The reciprocal integration, not incorporation, of the fringe into the indoors setting also provides insights for further study. While I initially started off in a theoretical dichotomy between the "slow space" of the hut and the "accelerated university" in the conference venue (Hanser, 2020a), the final conference of the tour had already unquestionably blurred these dividing lines. With critical self-reflection after each conference experience and experimenting with different ways of communicating the fringe to the indoors, the tiny house had been fully embraced by those representing the indoor conference. The organisers talked about the hut at the registration desk, made reference to it in announcements and all keynote speakers of the last conference turned up at the hut, blending into the ongoing talks of other delegates. Reciprocity seemed possible. Events are not only the playgrounds for established scientists to talk at, and for all others to talk about. Micro-events generate knowledges in their own right. Often, academic events are advertised merely as occasions for dissemination and knowledge exchange. There is a focus on a conference *about*, rather than a conference *for*. There are many potential ways forward for the transformational perspective on events research to generate novel ecologies *through* and *with* events, not only generating detached knowledge *about* them.

The experiment contributed insights that now need to be taken further; while every event inevitably creates an in-group and its own outsiders, a conference fringe can open up an experiential space for different sensory perceptions and visions for emancipatory conferencing spaces that emerge from lived experience of exclusion or non-recognition. While those categorised as the margins are not in the spotlight of the centre, a transient in-between can emerge which might reject the overwhelming exposure of the centre but which could at the same time claim a recognition that transcends the invisibility of the margins. Belongingness at conferences is then renegotiated in transit, overcoming invisibility as negative and exposure/visibility as positive features of conferencing.

This fluid methodology cannot provide a defining term to describe the transient struggles for place-making from the fringe, but it can speak through a visualisation of this movement (see Figure 13.10). *Fringe-making* alone would falsify the

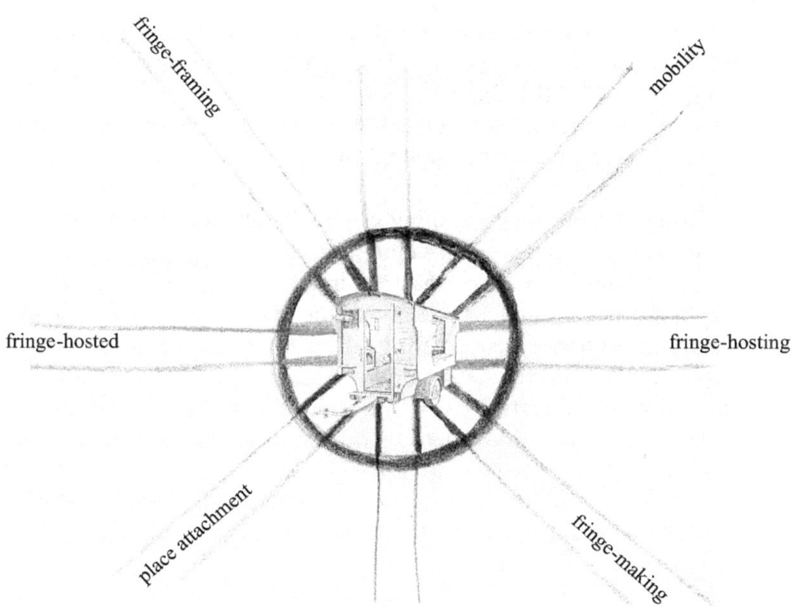

Figure 13.10 The centrality of movement between categories for an emancipatory fringe

Source: Author.

actual role of the researcher. A fringe is always already there, it does not need to be made by a researcher. It can only be made *more visible*. *Fringe-hosting* as a term also conveys an incomplete picture. While the researcher can be a host for delegates coming into the tiny house, the complexity of reciprocity needs to be investigated further. The fringe is equally hosted by the large-scale conferences and the gestures of welcome are embedded in a singular ecosystem of micro-hospitalities to macro-hospitalities at each conference venue. The fringe might navigate as a mobile place in a more-than-sedentary perception of invisibilities. The pre-existing tiny house has allowed to offer a fluid frame: the walls of a cosy hut held open a loosely tied set of largely invisible conference experiences that might not have connected without the framed invitation from the fringe. This frame did not lock up marginality, freeze exclusions, reify non-belonging or essentialise difference. As a mobile framing of provisional encounters at the fringes, it constantly shifts, recreates, disrupts and potentially transforms. Waiting for visitors allows to connect solitudes through a shared and deeply relational fringe-ness and singularity. It is a methodological shape that is difficult to pin down and impossible to master but which invites to research in the mode of constant becoming. It sketches out research designs that may emancipate from ableist triumph and the conquest over fragility through perfection.

Collective futures?

Holding open an unoccupied location requires precarious activist work, particularly if carried out as an individual project that emerged "by accident" and not through targeted research funding. This chapter has not intended to formulate a "fringe manual". Still, the long-term sustainability of conference fringes ought to be a major concern, based on the limitations of short-term provision for enabling support structures in the context of (dis-)ability. A major focus could be on those who will "perform well because of the social contexts and supports, and struggle when their enabling environments and services disappear" (Kapp, 2020, p. 309). As a solitary commitment, the approach has reached its expiry date. Continuing to run the space by myself would perpetuate a very tiring multitasking and paradoxically reinforce the overreliance on individualistic journeys that the relationalities of a fringe disrupt. To facilitate structural shifts in conferencing experiences, the methodological potential now requires to be situated in a novel perspective that moves from the singular to the plural: "to develop a sense of community is hard work and long term, especially in building levels of connectedness, belonging and support" (Jepson & Clarke, 2016, p. 4). The next step of possible tiny house conferencing methodologies requires a team with time commitment, shared on-site responsibilities and co-facilitating of improvised arts.

COVID-19 could initially seem like a challenge to this plan. However, in the journey of this fringe format something surprising frequently pops up to take the experiment in its whole emancipatory and relational "bulkiness" yet another tiny step further. An unexpected experience took place because of not despite the novel situation with COVID-19. At an Early Career Researcher forum at the University of Freiburg, my submitted proposal for a tiny house "dream tank" created a lot of interest among the organisers. Because the whole event had to be moved online, we decided to organise a one-day tiny house installation on the campus as a preview before the virtual event. While people could only enter the hut one by one because of the strict sanitary requirements, the organising experience allowed them to get a glimpse of a possible sustainable way to work. Where many conference organisers are usually too busy to enter the hut during the organising of the centre stages, I found myself running the in-person hut preview day together with the main organisers. I had become used to "delivering" the fringe independently of the main organisers. At many events, they were usually so busy in "delivering" the centre stage that they could not sit inside the hut. Here, we shared the tasks and the fringe brought a common on-site experience to the digitalised encounter. It is in unplanned, yet meaningful experiences like this pre-conference public day that the lines of indoors, outdoors and virtual conferencing formats are blurred to create new in-betweens.

Acknowledgements

This work was completed during a PhD scholarship funded by the SCDE Attainment Challenge Project.

I would like to highlight the inspiring input from two Early Career Researcher events. Discussions helped me considerably in the writing-up phase of this chapter:

German Tourism Research Colloquium July 2020 (AK Tourismusforschung/Deutschen Gesellschaft für Tourismuswissenschaft e.V.), Online

Emerging Scholars Forum of the Association for Canadian Studies in German-speaking Countries (GKS) September 2020, Freiburg and Online

References

Angner, E. (2020). Epistemic humility – Knowing your limits in a pandemic. *Behavioral Scientist*. Available from https://behavioralscientist.org/epistemic-humility-coronavirus-knowing-your-limits-in-a-pandemic/ [Accessed 28 August 2020].

Antchak, V. (2020). Marginalisation and events. *Annals of Leisure Research*. 23 (2), 264–266.

Augé, M. (1995). *Non-places: Introduction to an anthropology of supermodernity*. London: Verso.

Bochner, A.P., & Riggs, N.A. (2014). Practicing narrative inquiry. In: Leavy, P. (ed.) *The Oxford handbook of qualitative research*. Oxford: Oxford University Press, pp. 195–222.

Brown, N., & Leigh, J.S. (eds.) (2020). *Ableism in Academia: Theorising experiences of disabilities and chronic illnesses in higher education*. London: UCL Press.

Brown, N., Thompson, P., & Leigh, J. (2018). Making academia more accessible. *Journal of Perspectives in Applied Academic Practice*. 6 (2), 82–90.

Chapman, R. (2020). Neurodiversity, disability, well-being. In: Bertilsdotter Rosqvist, H., Chown, N., & Stenning, A. (eds.) *Neurodiversity studies: A new critical paradigm*. London: Routledge, pp. 57–72.

Conradson, D. (2011). The orchestration of feeling: Stillness, spirituality and places of retreat. In: Bissell, D., & Fuller, G. (eds.) *Stillness in a mobile world*. London: Routledge, pp. 71–86.

Conradson, D. (2016). Fostering student mental well-being through supportive learning communities. *Canadian Geographer*. 60 (2), 239–244.

Dashper, K., & Finkel, R. (2020). Accessibility, diversity, and inclusion in the UK meetings industry. *Journal of Convention & Event Tourism*. 1, 1–15.

Gillberg, C. (2020). The significance of crashing past gatekeepers of knowledge: Towards full participation of disabled scholars in ableist academic structures. In: Brown, N., & Leigh, J. (eds.) *Ableism in academia: Theorising experiences of disabilities and chronic illnesses in higher education*. London: UCL Press, pp. 11–30.

Hall, E., & Wilton, R. (2016). Towards a relational geography of disability. *Progress in Human Geography*. 41, 727–744.

Hanser, C.H. (2020a). Collective temporal activism as a gamechanger for the academy: Reframing conference hospitalities among colleagues under pressure. *International Review of Qualitative Research*, 1–18. DOI: 10.1177/1940844720968207 (online first).

Hanser, C.H. (2020b). Community development around the wood fire stove of a mobile tiny house: Experiencing the civic shelter of the welcome hut. *Practice Insights. The Community Development Journal of the IACD*. 16, 26–28.

Henderson, E.F., & Moreau, M.-P. (2020). Carefree conferences? Academics with caring responsibilities performing mobile academic subjectivities. *Gender and Education*. 32 (1), 70–85.

Hill Collins, P. (1999). Learning from the outsider within: The sociological significance of black feminist thought. *Social Problems*. 33 (6), 14–32.

Horton, J., & Tucker, F. (2014). Disabilities in academic workplaces: Experiences of human and physical geographers. *Transactions of the Institute of British Geographers*. 39 (1), 76–89.

Hughes, H., Franz, J., & Willis, J. (eds.) (2019). *School spaces for student well-being and learning: Insights from research and practice*. Cham: Springer.

Jazeel, T. (2019). Singularity. A manifesto for incomparable geographies. *Singapore Journal of Tropical Geography*. 40 (1), 5–21.

Jepson, A., & Clarke, A. (2016). An introduction to planning and managing communities, festivals and events. In: Jepson, A., & Clarke, A. (eds.) *Managing and developing communities, festivals and events*. London: Palgrave, pp. 3–15.

Judge, S.M. (2018). Languages of sensing: Bringing neurodiversity into more-than-human geography. *Environment and Planning D: Society and Space*. 36 (6), 1101–1119.

Kapp, S. (2020). Conclusion. In: Kapp, S. (ed.) *Autistic community and the neurodiversity movement stories from the frontline*. London: Palgrave Macmillan, pp. 305–318.

McFarlane, C. (2019). On singularity. *Singapore Journal of Tropical Geography*. 40 (1), 30–33.

Mittelmark, M.B., & Bauer, G.F. (2016). The meanings of salutogenesis. In: Mittelmark, M.B., Sagy, S., Eriksson, M. et al. (eds.) *The handbook of salutogenesis*. Cham: Springer, pp. 7–13.

Nelson, R. (2013). *Practice as research in the arts*. Basingstoke: Palgrave Macmillan.

Nicolson, D.J. (2017). *Academic conferences as neoliberal commodities*. Basingstoke: Palgrave Macmillan.

Parker, M., & Weik, E. (2014). Free spirits? The academic on the aeroplane. *Management Learning*. 45 (2), 167–181.

Russell, G. (2020). Critiques of the neurodiversity movement. In: Kapp, S. (ed.) *Autistic community and the neurodiversity movement*. Singapore: Palgrave Macmillan, pp. 287–303.

Santos, B.S. (2012). Public sphere and epistemologies of the south. *Africa Development*. XXXVII (1), 43–67.

Santos, B.S. (2018). *The end of the cognitive empire: The coming of age of epistemologies of the south*. London: Duke University Press.

Shahjahan, R.A. (2019). On 'being for others': Time and shame in the neoliberal academy. *Journal of Education Policy*. 5 (1), 1–27.

Spracklen, K., & Lamond, I.R. (2016). *Critical event studies*. London: Routledge.

Walters, T. (2019). A tripartite approach to accessibility, diversity and inclusion in academic conferences. In: Rebecca, F., Briony, S., & Majella, S. (eds.) *Accessibility, inclusion, and diversity in critical event studies*. London: Routledge, pp. 230–241.

Walters, T., & Jepson, A. (2019). Understanding the nexus of marginalisation and events. In: Walters, T., & Jepson, A. (eds.) *Marginalisation and events*. London: Routledge, pp. 1–16.

Young, I.M. (1990). *Justice and the politics of difference*. Princeton, NJ: Princeton University Press.

14 Lost in a crowd? Social isolation, loneliness, and events

Troy D. Glover

Introduction

Social isolation and loneliness pervade disturbingly across the world as increasingly serious and complex health problems. "Even by the most conservative estimates," wrote Holt-Lunstad (2020, p. 2), "loneliness affects one in five adults." As a result, its prevalence ought to be cause for serious concern, given that robust evidence from multiple meta-analyses and large-scale prospective epidemiological studies reveals social isolation and loneliness significantly increase the risk for morbidity and early mortality (Holt-Lunstad, Smith & Layton, 2010). Put simply, social connectedness – "the sense of belonging and subjective psychological bond that people feel in relation to individuals and groups of others" (Haslam *et al.*, 2015, p. 1) – matters significantly for our health and well-being and its absence can have dire consequences for those of us who feel a lack of connection to others (Pinker, 2014).

If things weren't bad enough, the emergence of the COVID-19 pandemic in 2020 resulted in workplace closures and the so-called "social distancing" measures that limited physical contact to mitigate the spread of infection, thereby threatening to inflame already high levels of loneliness. Physical restrictions adopted to "flatten the curve," while aimed paternalistically to reduce exposure to the virus, constrained mobility, restricted people to isolate at home (sometimes by themselves), and kept them from seeing their family and friends in person. Ironically, while encouraged to "socially distance," people needed more than ever to connect socially with others.

Interestingly, results about loneliness from research early in the pandemic were mixed. The COVID-19 Social Study, a research study run by University College London in the United Kingdom, reported that loneliness appeared worse than before COVID-19 with people from Black, Asian heritage, and minority ethnic groups, low-income households, individuals with mental illnesses, and young adults disproportionately at risk. However, other results from the United Kingdom revealed that loneliness in the first few weeks of the lockdown stabilised; although, for many people, these levels were high, to begin with and showed no signs of improvement (Bu *et al.*, 2020). Meanwhile, in the United States, loneliness appeared to level out, and in some cases, even improve (Killam, 2020).

And in Germany, preliminary evidence showed an initial increase, followed by a decrease in loneliness as people adapted and found ways to maintain social connection, despite the pandemic (Buecker *et al.*, 2020). All told, COVID-19 seemingly made people recognise the importance of their relationships, for they appeared to prioritise their social connections and became more aware and appreciative of the relationships in their lives.

Even with an ostensibly greater recognition of the importance of social connection that emerged from the pandemic, social isolation and loneliness continue to persist disconcertingly and globally as widespread health problems. Finding ways to help those battling these serious health matters warrants attention and creativity, especially in the age of the Novel Coronavirus. In my view, events offer a potential solution. While we know events can enhance social connectedness (Hassanli, Walters & Williamson, 2020; Stieler & Germelmann, 2016; Devine & Quinn, 2019), they are seldom considered for their strategic potential in addressing social isolation and loneliness. Accordingly, this chapter explores this potential and offers the field of event management direction in moving forward in combatting this challenging social problem.

An epidemic during a pandemic

Though often conflated, social isolation and loneliness refer to distinct concepts. Social isolation refers to "an objective situation, namely the actual absence of informal supportive relationships," whereas loneliness describes "a subjective and negatively experienced discrepancy between the quality and quantity of existing relationships and a person's desires or standards with regard to relationships" (Machielse, 2015, p. 340). Loneliness, in other words, can be thought of as *subjective* social isolation. Given these definitions, possibilities exist that individuals with many social connections can feel lonely, while others with few, if any, meaningful social connections can feel connected. Chances are, though, everyone has felt lonely at some point in his or her life. Loneliness produces social isolation when people, for whatever reason, withdraw from their social ties and emerges as a result of social isolation when people become disconnected from their sources of social interaction, namely their family, friends, neighbours, and co-workers.

To be sure, multiple studies provide evidence of the presence and prevalence of social isolation and loneliness (Holt-Lunstad *et al.*, 2015). The number of people with no confidants has evidently tripled over a 30-year period (McPherson, Smith-Lovin & Brashears, 2006), single-person households have increased (Klinenberg, 2016), few adults belong to local community or social groups (Holt-Lunstad, 2017; Pew Research Center, 2009), and volunteerism is on the decline (U.S. Department of Labor, Bureau of Statistics, 2016). These social trends, among others, put people at greater risk for social isolation and health problems. Feelings of loneliness, meanwhile, pervade disturbingly higher among those under the age of 25 and over the age of 65 (Victor & Yang, 2012). Irrespective of the age group, those with fewer supportive social ties score highest when measured for loneliness (Wilson & Moulton, 2010). Amazingly, Cigna (2020) reported that loneliness among Americans increased to 61% in 2019, up from 47% in 2018. Given the data, loneliness has emerged as a clear and present epidemic.

While the growth of social isolation and loneliness signals an alarming trend, understanding their impact on health and well-being is clear. The overwhelming evidence shows social connectedness affects health and well-being positively and significantly (Pinker, 2014). Positive relationships and social connections enable people to flourish. Having someone to rely upon during stressful life experiences allows individuals to cope effectively (Glover & Parry, 2008). People are better off when they believe they have and actually do interact regularly with others in their lives who care about them (Holt-Lunstad, 2017). If people feel lonely most or all of the time, their loneliness can influence their well-being and their ability to function effectively in society. Chronic loneliness leads to poor physical, mental, and personal health, and it can have adverse effects on the communities to which lonely people belong (Holt-Lunstad *et al.*, 2015). Ultimately, the quality of our relationships and friendships at home, work, and school, and in our communities matters tremendously. For all of these reasons, governments at all levels regard loneliness as an issue of increasing importance.

Though COVID-19 seemingly failed to exacerbate levels of social isolation and loneliness in any meaningful way (Killam, 2020), at least early on during the 2020 pandemic, public officials appeared to remain cognisant of, if not necessarily proactive about, the potential negative consequences of keeping people apart socially. The shift in language from "social distancing" to "physical distancing" early on recognised the necessity of the latter for safety and health purposes but rejected the former in terms of its misguided messaging to the public. People need social connection for health purposes. Early efforts to close parks and playgrounds, cancel events, and discourage social gatherings ran counter to recommendations on how to reduce loneliness but seemed necessary given the aim to "flatten the curve" of the disease. Moving forward, especially in response to a "subsequent wave" or future pandemics, communities will need to (re)consider ways they can support safe social interaction. Elevating the importance of social connectedness in the public realm, as well as the public *health* realm, through a variety of community efforts holds the potential to significantly improve population health (Holt-Lunstad, Robles & Sbarra, 2017). In my view, events represent one area of consideration because of their potential for strengthening social connectedness.

Events and social connectedness

Though physical distancing may be the right response during a pandemic, people unquestionably need social connection to counteract the loneliness epidemic. Events reinforce social connectedness as spaces in which the human need for "frequent, affectively pleasant interactions with others" (Jaeger & Mykletun, 2013, p. 214) can be satisfied. In this sense, events represent an important and promising "sphere of sociability" (Glover & Parry, 2008, p. 221) or "site of civic promise" (Amin, 2006, p. 1020). When attended in-person, they create shared rhythms among participants in a physical space, albeit by interrupting, disrupting, or challenging everyday uses of the space to facilitate novel experiences (Glover, 2020) and putting attendees in touch "with something exceptional" beyond themselves (Taylor, 2007, p. 482). Events attract participants by enabling attendees to deviate from their daily routines to experience something fresh and different from everyday life.

For people who feel socially isolated, who feel lonely, events offer potentially inviting spaces to engage with others and foster, if anything, minimal social connections to other people or groups (Walton *et al.*, 2012). While Walton *et al.* (2012) referred to this form of connection as "mere belonging," Horgan (2012, p. 619) described it as "soft solidarity," a form of mutuality implicit in relations among strangers without the requirement of explicit recognition. Because it is impossible to know everyone, anonymity necessarily characterises our everyday experiences in modern urban life (Hirschauer, 2005). Being "co-present" in public – that is, remaining "physically proximate, but socially distant" (Horgan, 2012, p. 608) – normalises strangeness among people (Hirschauer, 2005), which has the potential to contribute to social isolation and feelings of loneliness. In-person events, however, not only put people in close physical proximity to each other, but also work to bridge social distance among attendees so that, even if only for a brief moment in time (during the course of an event), attendees feel connected as familiar strangers – people who are recognisable, but with whom one interacts minimally.

Familiar strangers are not trivial relations. Welcoming a social interaction for a brief, but authentic, moment establishes a temporary bond (Campos-Castillo & Hitlin, 2013) and expands our imagined (Glover, Macnaughton & Mock, 2021) and psychological sense of community (Hassanli *et al.*, 2020). What Herzfeld (2014, p. 2) described as "cultural intimacy," in this case produced by events, transforms the abstract idea of community into intimate expressions of felt solidarities. Accordingly, an imagined community inheres in the interactions and experiences with other event attendees to reinforce a collective identity derived from an acknowledgement of affinity. Dissemination of group symbols at events, moreover, emphasises community boundaries and enhances perceived collective unity. These reinforcing experiences serve to build connections, solidarity, and identity (Glover, Macnaughton & Mock, 2021).

When attending events, most attendees seek the most basic of social interaction: the desire to avoid being alone. "Mingling with the crowd," wrote Torres, Moreira and Lopes (2018, p. 318), "is an antidote to daily isolation and works through living the events in the company of others." People who attend alone connect with the fleeting and nameless crowd on some level to reinforce or develop their sense of belonging. "The resultant social renewal," added Torres, Moreira and Lopes (2018, p. 318), "strengthen[s] the social belonging of each individual and compensate[s] for traits of contemporary society, such as loneliness and competition."

In many ways, events have the potential to create temporary comfort zones. They foster "a sense of feeling at ease, or at home in a world which can appear external and threatening" (Dupuis & Thorns, 1998, p. 30). This sense of "ontological security" offers an escape from loneliness. If only momentarily, attendees feel connected to others – part of something bigger than themselves. Events change the character of the places where they occur and create more permissible situations for strangers to talk to each other (Simões Aelbrecht, 2016). Though not overwhelming, Wilks (2011) found evidence of temporary connections among event participants who chatted with adjacent audience members at festivals, often during refreshment breaks, even if only to be friendly. A general trust in event attendees and a feeling of safety at the event enabled these participants to feel comfortable making casual conversation with others.

Conversation is not always necessary, though. Simply observing or being close to social interactions in a public space is enough for people to report feeling a greater sense of happiness, safety, and comfort with their community (Cattell *et al.*, 2008). Some attendees may prefer to simply "watch the world go by," an activity Woolley (2003, p. 20) described as "passive recreation," to derive pleasure from the event. In this sense, an event has the potential to be a less intense social environment insofar as direct social interaction may be limited or perhaps not required of attendees at all. What Hochschild (1983, p. 7) called emotional labour, which "requires one to induce or suppress feeling in order to sustain the outward countenance that produces the proper state of mind in others," may be subdued in such circumstances. Consequently, entry into an event can be non-threatening and potentially inviting.

Even so, mutual emotion can occur without conscious awareness, likely through nonconscious mimicry of the expressions of others that eventually leads attendees to experience the same emotion (Campos-Castillo & Hitlin, 2013). This emotional contagion is particularly evident during the phenomenon known as collective effervescence, a powerful psychological experience connected to strong emotions and well-being when people are engaged in collective activities (Durkheim, 1912; Gabriel *et al.*, 2020). Citing what they called a "hive hypothesis," Haidt and Kesebir (2008, p. S136) argued people must lose themselves occasionally in a larger social experience to achieve higher levels of individual well-being and protect themselves against what Cacioppo *et al.* (2015, p. 239) referred to as "collective loneliness." Accordingly, they suggested activities, such as events, that enable collective effervescence are more effective in increasing well-being and decreasing loneliness. Ultimately, events can have a direct influence on loneliness and well-being by raising people's spirits and making them feel as though they belong.

Online events when in-person is not possible

The face-to-face benefits of social connection are clear (Pinker, 2014). However, COVID-19 and future pandemics necessarily require restricting the size of crowds and gatherings through physical distancing measures to avoid the spread of infection (Honey-Rosés *et al.*, 2020), which makes in-person events challenging if not impossible to host. While events contribute to a vibrant social and cultural life, they tend to put people in close proximity and, therefore, at risk of contagion. People may understandably feel isolated during a pandemic, but fortunately the means, if not confidence, exists for them to assert control over their social lives and their social health to avoid feeling lonely. Organising or attending virtual gatherings offers them one possibility in this regard (Killam, 2020).

In the 21st century, people can be creative with digital tools in ways that were simply unimaginable to previous generations. Undoubtedly, the multitude of video hosting services (e.g., Zoom, Microsoft Teams) and video chat apps (e.g., WhatsApp, FaceTime) makes it easier than ever to replicate social gatherings, thereby allowing people to invent new ways of coming together during and after the pandemic. Indeed, in response to physical distancing declarations, people around the world hosted virtual church services, virtual dance parties and raves, virtual happy hours, and virtual potlucks, among other virtual events (Palamar & Acosta, 2020). Netflix, the online streaming service, introduced Netflix Party, a feature that allows users to link up with friends and host long-distance movie nights. Book clubs, choirs, and fitness classes were all shifted online with conducive software. Even massive online gaming platforms, such as Fortnite and Roblox, became virtual hangouts for young people. In so many ways, communications technology helped us avoid a serious social decline during this difficult period. With this in mind, Davies (2020) predicted growth in the number and importance of virtual or online events in the aftermath of the pandemic.

Turning to digital technology to reduce loneliness among vulnerable and hard-to-reach people during a pandemic warrants consideration by event management professionals. Research by Tsai *et al.* (2010) and Blazun, Saranto and Rissanen (2012) shows videoconference and online social networking can be successful in reducing the experience of loneliness. Features of online communication seem to be particularly appealing to lonely and/or socially anxious individuals insofar as online and virtual platforms can be perceived as "protected" environments, wherein participants feel they can express their inner selves. To be sure, virtual gatherings offer potentially accessible, inclusive, flexible and responsive spaces to connect, and places where lonely people can locate themselves and be met "as [they] are" (Black *et al.*, 2020, p. 126). Ultimately, virtual gatherings allow users to fulfil their needs of social interaction, self-disclosure, and identity exploration in a less threatening context. For virtual events to achieve these outcomes, however, they need to be planned, organised, and implemented deliberately to ensure inclusion and belonging (see more later).

Despite their benefits during a time of physical distancing, virtual events are by no means a panacea. To be sure, events themselves, whether in-person or virtual, can be isolating. Social interactions at events can stimulate negative reactions, and exposure to some participants can contribute towards feelings of insecurity and fear (Nordvall *et al.*, 2014). Social presence theory suggests that online communication deprives people of the sense that another person is participating in social interactions, thereby keeping contacts impersonal (Tu & McIsaac, 2002).

Moreover, passive coping strategies by lonely people that incorporate solitary participation, such as watching screens passively, consuming entertainment submissively, or spending time on a computer or mobile device without engaging meaningfully with others can further isolate individuals (Toepoel, 2013). Putnam (2000) cited the privatisation and individualisation of our use of leisure time as disruptions to our ability to forge meaningful relationships and build social capital.

Feeling lonely in an age of hyperconnectivity may seem counterintuitive because virtual platforms make it undeniably easier for people to connect socially with others. Nevertheless, advances in digital technologies appear to enhance social connectivity at the expense of the depth of our relationships (Parigi & Henson, 2014). Expanding our number of social ties through virtual events may not necessarily generate more meaningful connections, but rather add to the cognitive and emotional costs of maintaining and sustaining those connections, to the point of creating or exacerbating a sense of isolation (Glover, 2018). Virtual events may make it possible for one to connect with a seemingly boundless number of people online while remaining subjectively socially isolated oneself.

By shifting events online, furthermore, event organisers risk isolating "those who do not have internet access, do not know how to use videoconferencing applications, or who cannot afford internet-enabled devices or an internet plan" (Hartt, 2020, p. 2). In addition, older adults, the people who belong to the same age cohort most likely to be widowed and most susceptible to infection during a pandemic, are least likely to be comfortable or competent users of new apps and platforms (Hartt, 2020). Consequently, the shift to virtual events can further isolate the least connected and most vulnerable. Recognition of these possibilities needs to guide the development of thoughtful and inclusive virtual events if the event field hopes to engage lonely people effectively.

Events as interventions

People themselves dramatically underestimate the importance of social factors for mortality (Haslam *et al.*, 2018) but so do event management professionals. Though social connectedness may be fundamental to the mission of many events, utilising events as interventions to address social isolation and loneliness explicitly warrants more attention from the events field. The pandemic ought to show how the mission of event management should evolve to include serving as a provider of social and health services, defined broadly, in the effort to rebuild and reconnect communities.

Unfortunately, the truth of the matter is that people rarely go alone to any type of event (Torres, Moreira & Lopes, 2018) and bridging-type social interactions between people previously unknown to each other are, frankly, uncommon at events (Wilks, 2011). People who deal with loneliness tend to adopt either active or passive coping strategies (Vanhalst *et al.*, 2012), which can influence whether they attend. Active coping strategies involve action to alleviate or resolve loneliness by regarding it, in a positive way, as a challenge with the expectation of future inclusion and social exchanges. Passive strategies, by contrast, overestimate the potential social threats, elicit feelings of discouragement that prevent individuals from reaching out, and ultimately fail to deal with the problem directly. Put another way, *loneliness reduction* suggests that an unfulfilled need drives individuals to pursue solutions to resolve their loneliness, whereas *loneliness perpetuation* desensitises the person to any anticipation of potential benefits from social inclusion, so their unmet need to belong remains unsatisfied (see Vanhalst *et al.*, 2012). If event managers are going to address loneliness, they need to recognise these strategies and consider deliberate ways to encourage participation among this target group.

Gardiner, Geldenhuys and Gott (2018) identified three key common characteristics of effective interventions to address loneliness that event professionals ought to consider. First, they noted successful interventions require *adaptability* to a local context. In other words, events, whether virtual or in-person, need to be regarded as locally relevant/authentic to bring lonely participants closer to their imagined communities and to encourage their participation. Jepson and Clarke (2013, p. 8) noted events "are susceptible to a system of cultural production that aims to make the [event] product as widely appealing as possible and in doing so can change it to a more homogenous or commodified production, which is then disconnected from the local communities it set out to serve." Accordingly, they underscored the importance of the local community in the success of any event. Adaptability, put simply, amounts to paying attention to the spirit of community. This approach especially applies to events targeting lonely people.

Second, involving beneficiaries of the intervention in its design and implementation by using a community development approach enhances the likelihood of positive outcomes for lonely people. Piazzi and Harris (2016) listed several benefits to event organisers who engage in such practices, including overcoming resistance or opposition to an event within a host community; avoiding decisions that may otherwise cause conflict; and enhancing access to potential volunteers. Adopting a community development approach lines up well with the idea that events ought to be "for the people by the people." Interestingly, introverts are often thoughtful resources upon which to draw when thinking through events and gatherings because they can speak to features of events that constrain participants for people like themselves (Parker, 2020). Engaging people who are lonely in the design and implementation of events makes for more inviting events.

Third, activities or interventions that support productive engagement are more successful in alleviating loneliness than those with no explicit goal or purpose. Parker (2018) argued those who organise events ought to think more intentionally about the purpose of their gatherings. She suggested a set of what she called "spiritual" questions: why are we doing this in the first place? Is it really needed? Who is this for? And who gets to decide? These questions, she explained, invite a set of questions that every gathering needs to be answered: what do we need at this moment and how might we gather around that? If the event is intended to address social isolation and loneliness, event planners must factor this aspiration into the design, planning, and implementation deliberately.

Conclusion

Whether under conditions of a pandemic or not, social isolation and loneliness warrant attention because of their severe assault on health and well-being. Events offer a potential solution to this wicked problem by inviting people to join forums that welcome greater social connectedness and belonging, outcomes that enable people to thrive in life. Irrespective of whether an event is offered on or offline, opportunities exist for events to play this important role in a society in which loneliness pervades as an epidemic. The event field will no doubt expand its social relevance by serving such a purpose.

References

Amin, A. (2006). The good city. *Urban Studies*. 43 (5–6), 1009–1023.

Black, A.L., Crimmins, G., Dwyer, R., & Lister, V. (2020). Engendering belonging: Thoughtful gatherings with/in online and virtual spaces. *Gender and Education*. 32 (1), 115–129.

Blazun, H., Saranto, K., & Rissanen, S. (2012). Impact of computer training courses on reduction of loneliness of older people in Finland and Slovenia. *Computers in Human Behavior*. 28 (4), 1202–1212.

Bu, F., Steptoe, A., Mak, H.W., & Fancourt, D. (2020). Time-use and mental health during the COVID-19 pandemic: A panel analysis of 55,204 adults followed across 11 weeks of lockdown in the UK. *medRxiv*. http://doi.org/10.1101/2020.08.18.20177345.

Buecker, S., Horstmann, K.T., Krasko, J., Kritzler, S., Terwiel, S., Kaiser, T., & Luhmann, M. (2020). Changes in daily loneliness during the first four weeks of the Covid-19 lockdown in Germany. https://doi.org/10.1016/j.socscimed.2020.113541.

Cacioppo, S., Grippo, A.J., London, S., Goossens, L., & Cacioppo, J.T. (2015). Loneliness: Clinical import and interventions. *Perspectives on Psychological Science*. 10 (2), 238–249.

Campos-Castillo, C., & Hitlin, S. (2013). Copresence: Revisiting a building block for social interaction theories. *Sociological Theory*. 31 (2), 168–192.

Cattell, V., Dines, N., Gesler, W., & Curtis, S. (2008). Mingling, observing, and lingering: Everyday public spaces and their implications for well-being and social relations. *Health & Place*. 14 (3), 544–561.

Cigna. (2020). *Loneliness and the workplace* [press release]. Available from www.cigna.com/static/www-cigna-com/docs/about-us/newsroom/studies-and-reports/combatting-loneliness/cigna-2020-loneliness-factsheet.pdf.

Davies, K. (2020). Festivals post Covid-19. *Leisure Sciences*. DOI: 10.1080/01490400.2020.1774000.

Devine, A., & Quinn, B. (2019). Building social capital in a divided city: The potential of events. *Journal of Sustainable Tourism*. 27 (10), 1495–1512.

Dupuis, A., & Thorns, D.C. (1998). Home, home ownership and the search for ontological security. *The Sociological Review*. 46 (1), 24–47.

Durkheim, E. (1912). *The elementary forms of the religious life*. New York: The Free Press.

Gabriel, S., Naidu, E., Paravati, E., Morrison, C.D., & Gainey, K. (2020). Creating the sacred from the profane: Collective effervescence and everyday activities. *The Journal of Positive Psychology*. 15 (1), 129–154.

Gardiner, C., Geldenhuys, G., & Gott, M. (2018). Interventions to reduce social isolation and loneliness among older people: An integrative review. *Health & Social Care in the Community*. 26 (2), 147–157.

Glover, T.D. (2018). All the lonely people: Social isolation and the promise and pitfalls of leisure. *Leisure Sciences*. 40 (1–2), 25–35.

Glover, T.D. (2020). Bringing public spaces to life: The animation of public space. In: Mehta, V., & Palazzo, D. (eds.) *The Routledge companion to public space*. New York: Routledge, pp. 414–425.

Glover, T.D., Macnaughton, J.F.P., & Mock, S.E. (2021). Inside out: The role(s) of leisure in the endogenous and exogenous pathways to social capital. In: Glover, T.D., & Sharpe, E. (eds.) *Leisure communities: Rethinking mutuality, collective identity and belonging in the new century*. New York: Routledge, pp. 159–169.

Glover, T.D., & Parry, D.C. (2008). Friendships developed subsequent to a stressful life event: The interplay of leisure, social capital, and health. *Journal of Leisure Research*. 40 (2), 208–230.

Haidt, S., & Kesebir. (2008). Hive psychology, happiness, and public policy. *The Journal of Legal Studies*. 37, 156.

Hartt, M. (2020). COVID-19: A lonely pandemic. *Cities & Health*, 1–3.

Haslam, C., Cruwys, T., Haslam, S.A., & Jetten, J. (2015). Social connectedness and health. In: Pachana, N.A. (ed.) *Encyclopedia of geropsychology*. Singapore: Springer, pp. 2174–2182.

Haslam, S.A., McMahon, C., Cruwys, T., Haslam, C., Jetten, J., & Steffens, N.K. (2018). Social cure, what social cure? The propensity to underestimate the importance of social factors for health. *Social Science & Medicine*. 198, 14–21.

Hassanli, N., Walters, T., & Williamson, J. (2020). 'You feel you're not alone': How multicultural festivals foster social sustainability through multiple psychological sense of community. *Journal of Sustainable Tourism*. DOI: 10.1080/09669582.2020.1797756.

Herzfeld, M. (2014). *Cultural intimacy: Social poetics in the nation-state*. New York: Routledge.

Hirschauer, S. (2005). On doing being a stranger: The practical constitution of civil inattention. *Journal for the Theory of Social Behaviour*. 35 (1), 41–67.

Hochschild, A. (1983). *The managed heart: Commercialization of human feeling*. Berkeley, CA: University of California Press.

Holt-Lunstad, J. (2017). The potential public health relevance of social isolation and loneliness: Prevalence, epidemiology, and risk factors. *Public Policy & Aging Report*. 27 (4), 127–130.

Holt-Lunstad, J. (2020, June 22). Social isolation and health. *Health Affairs Policy Brief*. DOI: 10.1377/hpb20200622.253235.

Holt-Lunstad, J., Robles, T.F., & Sbarra, D.A. (2017). Advancing social connection as a public health priority in the United States. *American Psychologist*. 72 (6), 517–530.

Holt-Lunstad, J., Smith, T.B., Baker, M., Harris, T., & Stephenson, D. (2015). Loneliness and social isolation as risk factors for mortality. *Perspectives on Psychological Science*. 10 (2), 227–237.

Holt-Lunstad, J., Smith, T.B., & Layton, J.B. (2010). Social relationships and mortality risk: A meta-analytic review. *PLoS Medicine*. 7 (7), e1000316.

Honey-Rosés, J., Anguelovski, I., Chireh, V.K., Daher, C., Konijnendijk van den Bosch, C., Litt, J.S., . . ., Sánchez, U. (2020). The impact of COVID-19 on public space: An early review of the emerging questions – design, perceptions and inequities. *Cities & Health*, 1–17.

Horgan, M. (2012). Strangers and strangership. *Journal of Intercultural Studies*. 33 (6), 607–622.

Jaeger, K., & Mykletun, R.J. (2013). Festivals, identities, and belonging. *Event Management*. 17 (3), 213–226.

Jepson, A., & Clarke, A. (2013). Events and community development. In: Finkel, R., McGillivray, D., McPherson, G., & Robinson, P. (eds.) *Research themes for events*. Wallingford, OX: Cabi, pp. 6–17.

Killam, K. (2020, August 18). In the midst of the pandemic, loneliness has leveled out. *Scientific American*. Available from https://www.scientificamerican.com/article/in-the-midst-of-the-pandemic-loneliness-has-leveled-out/.

Klinenberg, E. (2016). Social isolation, loneliness, and living alone: Identifying the risks for public health. *American Journal of Public Health*. 106 (5), 786.

Machielse, A. (2015). The heterogeneity of socially isolated older adults: A social isolation typology. *Journal of Gerontological Social Work*. 58 (4), 338–356.

McPherson, M., Smith-Lovin, L., & Brashears, M.E. (2006). Social isolation in America: Changes in core discussion networks over two decades. *American Sociological Review*. 71 (3), 353–375.

Nordvall, A., Pettersson, R., Svensson, B., & Brown, S. (2014). Designing events for social interaction. *Event Management*. 18 (2), 127–140.

Palamar, J.J., & Acosta, P. (2020). Virtual raves and happy hours during COVID-19: New drug use contexts for electronic dance music partygoers. *International Journal of Drug Policy*. DOI: 10.1016/j.drugpo.2020.102904.

Parigi, P., & Henson, W. (2014). Social isolation in America. *Annual Review of Sociology*. 40, 153–171.

Parker, P. (2018). *The art of gathering: How we meet and why it matters*. New York: Riverhead Books.

Parker, P. (2020, March 17). How to be together apart in the time of coronavirus. *New York Times*, 1224 words.

Pew Research Center. (2009). *Social isolation and new technology: How the internet and mobile phones impact Americans' social networks*. Pew Internet & American Life Project. Available from www.pewinternet.org/2009/11/04/social-isolation-and-new-technology-2/.

Piazzi, F., & Harris, R. (2016). Community engagement and public events: The case of Australian folk festivals. *Event Management*. 20 (3), 395–408.

Pinker, S. (2014). *The village effect: How face-to-face contact can make us healthier and happier*. Toronto: Random House Canada.

Putnam, R.D. (2000). *Bowling alone: The collapse and revival of American community*. New York, NY: Simon & Schuster.

Simões Aelbrecht, P. (2016). 'Fourth places': The contemporary public settings for informal social interaction among strangers. *Journal of Urban Design*. 21 (1), 124–152.

Stieler, M., & Germelmann, C.C. (2016). The ties that bind us: Feelings of social connectedness in socio-emotional experiences. *Journal of Consumer Marketing*. 33 (6), 397–407.

Taylor, C. (2007). *A secular age*. Cambridge, MA: Harvard University Press.

Toepoel, V. (2013). Ageing, leisure, and social connectedness: How could leisure help reduce social isolation of older people? *Social Indicators Research*. 113 (1), 355–372.

Torres, E.C., Moreira, S., & Lopes, R.C. (2018). Understanding how and why people participate in crowd events. *Social Science Information*. 57 (2), 304–321.

Tsai, H.H., Tsai, Y.F., Wang, H.H., Chang, Y.C., & Chu, H.H. (2010). Videoconference program enhances social support, loneliness, and depressive status of elderly nursing home residents. *Aging & Mental Health*. 14 (8), 947–954.

Tu, C.H., & McIsaac, M. (2002). The relationship of social presence and interaction in online classes. *The American Journal of Distance Education*. 16 (3), 131–150.

U.S. Department of Labor, Bureau of Statistics. (2016). *Volunteering in the United States, 2015*. Available from www.bls.gov/news.release/volun.nr0.htm.

Vanhalst, J., Luyckx, K., Teppers, E., & Goossens, L. (2012). Disentangling the longitudinal relation between loneliness and depressive symptoms: Prospective effects and the intervening role of coping. *Journal of Social and Clinical Psychology*. 31 (8), 810–834.

Victor, C.R., & Yang, K. (2012). The prevalence of loneliness among adults: A case study of the United Kingdom. *The Journal of Psychology*. 146 (1–2), 85–104.

Walton, G.M., Cohen, G.L., Cwir, D., & Spencer, S.J. (2012). Mere belonging: The power of social connections. *Journal of Personality and Social Psychology*. 102 (3), 513.

Wilson, C., & Moulton, B. (2010). *Loneliness among older adults: A national survey of adults 45+*. Washington, DC: American Association of Retired Persons.

Wilks, L. (2011). Bridging and bonding: Social capital at music festivals. *Journal of Policy Research in Tourism, Leisure and Events*. 3 (3), 281–297.

Woolley, H. (2003). *Urban Open Spaces*. London: Spoon Press.

Index

For Product Safety Concerns and Information please contact our EU representative GPSR@taylorandfrancis.com
Taylor & Francis Verlag GmbH, Kaufingerstraße 24, 80331 München, Germany

www.ingramcontent.com/pod-product-compliance
Lightning Source LLC
LaVergne TN
LVHW021122300926
849160LV00010B/820
9781032011769